Fast Circuit Boards

Fast Circuit Boards

Energy Management

Ralph Morrison

This edition first published 2018

Registered Offices
John Wiley & Sons, Inc., 111 River Street, Hoboken, NJ 07030, USA

Editorial Office
111 River Street, Hoboken, NJ 07030, USA

For details of our global editorial offices, customer services, and more information about Wiley products visit us at www.wiley.com.

Wiley also publishes its books in a variety of electronic formats and by print-on-demand. Some content that appears in standard print versions of this book may not be available in other formats.

Library of Congress Cataloging-in-Publication Data

Names: Morrison, Ralph, author.
Title: Fast circuit boards : energy management / by Ralph Morrison.
Description: Hoboken, NJ : John Wiley & Sons, 2018. | Includes index. |
Identifiers: LCCN 2017037713 (print) | LCCN 2017046159 (ebook) |
ISBN 9781119413929 (pdf) | ISBN 9781119413998 (epub) | ISBN 9781119413905 (cloth)
Subjects: LCSH: Very high speed integrated circuits–Design and construction. |
Logic design.
Classification: LCC TK7874.7 (ebook) | LCC TK7874.7. M67 2018 (print) |
DDC 621.31–dc23
LC record available at https://lccn.loc.gov/2017037713

Cover Design: Wiley
Cover Image: © troyek/Getty Images

Set in 10/12pt Warnock by SPi Global, Pondicherry, India

Printed in the United States of America

10 9 8 7 6 5 4 3 2 1

Contents

Preface

If you are reading this preface you are probably involved in designing and laying out logic circuit boards. I have a story to tell you which you will not find on the internet or in other books. What I have to say has been put to practice and it works. It is not complicated but it is different. In this book, I ask you to go back to the basics so that I can explain the future. I hope you are willing to put forth the effort to go down this path.

I would like to thank my wife Elizabeth for her encouragement and help. She never complained when I spent days on end at my computer writing and rewriting. It takes a lot of dedicated time to write a book.

I would like to thank Dan Beeker of NXP Semiconductors. He is a principal engineer in Automotive Field Engineering. I have given many seminars arranged by Dan over the years. Using the material in my seminars he has been very effective in helping designers avoid problems. His experiences are proof that the material in this book, when put to practice, really works. His success has spurred me on. Highlights of this understanding are blocked out in the text as **Insights**.

This book presents some ideas that I have not seen in print or heard at conferences. I know that this does not prove that these ideas are new or novel. It could mean that I have not talked to the right people. My contact with engineers tells me they mainly come out of the same molds in school. The basic math and physics that is taught revolves around differential equations that in most cases solve problems using numerical techniques.

Computers work well in antenna design and in moving energy in wave guides. For a long time the problem of wiring circuit boards has been considered trivial and has not received very much attention. One of the reasons is that people have been getting by. That is no longer the case and it is time for a change. A big part of the problem is that sine waves and antenna or microwave design methods are not a fit for transmission lines on circuit boards where step functions, delays, and reflections take center stage.

To whet your appetite here are a few ideas that are treated in this book:

1) Logic is the movement of energy
2) Not all waves carry energy
3) We cannot measure moving field energy directly
4) Waves deposit, convert and move electric and magnetic field energy on transmission lines
5) Radiation only occurs on leading edges
6) Energy in motion is half electric and half magnetic
7) Via positioning controls radiation
8) Transmission lines can oscillate
9) Waves can convert stored electric field energy to stored magnetic field energy
10) Waves can convert stored magnetic field energy to stored electric field energy
11) Waves can convert stored energy into moving energy

In my career I have written 14 books, all published by John Wiley &Sons. I am 92 and I have been retired for some 25 years. That has not stopped me from giving seminars, doing consulting, and writing books. I often reflect on what keeps my writing and how I seem to be almost singular in my approach to interference issues. A lot has to do with the opportunities given to me in my career. Since this will probably be my last book I thought this would be an opportunity to provide the readers with some of my personal background. A lot of people have helped me over the years and my story is unique.

I was born in Highland Park, a suburb of Los Angeles, California on January 4, 1925 to immigrant parents who had no understanding or interest in science. I grew up in the great depression of the thirties when a cup of coffee was 5 cents. Cellophane and zippers were not a part of life. The last horse-drawn carriages brought fresh vegetables to our street. There was the ice man and houses had ice boxes. Raw or pasteurized milk was delivered in bottles by the milkman before I got up. Radios blared soap operas all day.

My early experiences with things electrical were crystal sets, radios, and building an audio amplifier. I learned how to measure voltage and calculate current flow. I used an oscilloscope in school to observe circuit voltages. I observed magnetics in terms of loud speakers, motors, and transformers. I formed images of current flow and voltage patterns. I knew about radio transmission and antennas from my amateur radio friends, but this area was a mystery to me. It was not until I entered college that I was introduced to electromagnetic fields. By then I had enough mathematics to work a few simple problems but my understanding of the electrical world was still very limited.

I started playing violin at age 4½ and my father got me a scholarship. I walked a mile to elementary school and I remember the Maypole in the playground. I walked a mile over a hill to Eagle Rock High School where I had my first brush with geometry, algebra, physics, and electric shop. I had some fine teachers. Ben Culley, one of my math teachers, went on to be dean of men at Occidental College. We had a radio at home and that intrigued me. I pestered the local radio repair shop and was allowed to help out by testing tubes. We had no automobile but I made the effort to bicycle to the Friday night lectures at Caltech. I saw the 100-inch Mount Wilson telescope lens when it was moved out of the optics lab. I saw the demonstration at the Kellogg high voltage lab. In my teen years I was leaning toward things electrical. Then Pearl Harbor was bombed. I remember Roosevelt's famous "infamy" radio speech. In 3 weeks I turned 17. I remember the air raid sirens and the blackouts. I remember when the Japanese shelled the west coast and the searchlights came on. I remember gas masks were issued and there was gas and food rationing. Members of my class were volunteering into the services and big changes were taking place in the lives around me.

I was drafted into the army in the April of 1943 and did basic training in Fresno, California. The army sent me to Oregon State College as part of an Army Specialized Training Program. I had my first bus and train ride. At OSC I had a few basic engineering courses. Much of the class material was a review for me. It was decided that the war was not going to last decades and the education of future engineers was not a high priority. My start in college lasted about 6 months and I was shipped off to the 89th infantry division at Hunter Liggett Military Reservation in California. I was given a course in radio repair at Fort Benning, Georgia. The division eventually ended up crossing Germany in Patton's third army. I saw bombed out cities. I watched and heard the bombing during the Rhine river crossing. I did my calculus through the University of California correspondence course in this period. I remember working problems when one of our own aircrafts was shot down because he was firing on us. The army went as far as Zwickau and I saw the exodus of slave labor. They were walking back to their home cities with no food or belongings. Our division uncovered one of the concentration camps. This was the first knowledge I had of what had been going on in Europe in the preceding years. It was hard to grasp.

After hostilities, it took months before we could return home. The war was still in progress in the Pacific. This meant that there were only so many ships available. I visited London, Brussels, Edinburgh, and Glasgow on passes. I saw the Loch Ness. Since I was a violinist, I took the opportunity to join a GI symphony orchestra. I spent a month in Paris before we were moved to Frankfurt. It was a big transition—from army life in war to a home in a French mansion. The orchestra toured Germany and Austria entertaining troops. I saw a lot of Europe including the war crimes trial at Nuremburg and the inside of Hitler's bunker in Berlin. We played in the Schonbrunn Palace in Vienna, in the Wagner Festspiel Haus in Beirut, and in Garmisch Partenkirchen. For a kid that had never left home, I had quite an adventure in the army.

Three years in the service and I finally returned home. I wanted to use the GI bill to get a college degree at Caltech. I was 21 years old. The first thing I did personally was add a room to our home so that I had a place to study. The backlog of students trying to continue their education was long and my only option was to take the junior entrance exam. I was given credit for my classes at Oregon State and my correspondence course. I studied all summer so that I could take exams in English, math, chemistry, and physics. I was one of six that was accepted. I chose physics as my major as I really did not know what direction to take. I finished my senior year without enough credits to graduate. I came back for two more terms and nearly finished all the courses needed to get a Masters in EE. I had used up my GI bill. I had a difficult time starting as a junior but somehow I made it. I graduated with the class of 1949. I still remember that my first physics course was given by Dr. Carl Anderson, the Nobel prize winner that discovered the positron. I was in a different world. I had no more funds and I had to go to work.

My first job was with a company called Applied Physics Corporation in Pasadena, CA. I worked for George W. Downs, a respected technical consultant with ties to Caltech, the US Navy, and the Atomic Energy Commission. My first assignment was to build a dc amplifier for Douglas Aircraft. In 1950, there were only vacuum tubes for gain. I was shown a circuit that used a mechanical chopper to stabilize a feedback amplifier. Vacuum tubes require hundreds of volts to operate and there had to be transformers to isolate the circuits from utility ground. My first dc amplifier and power supply weighed over 70 pounds. Looking back, it is hard to believe that this was progress. Based on what had been learned, my next dc amplifiers were much smaller and a group of amplifiers shared one power supply. These early designs were bought by the aircraft companies to handle the signals from strain gauges and thermocouples. These were the days when the first jet aircraft was still on the drawing board. I was given the job of building four analog computers that paralleled work done at Caltech. Each of these computers filled a room and used the dc amplifier designs I had worked on. These computers were sold to Lockheed, Douglas, and North American Aircraft.

Digital computing was in its infancy. When my first boss was asked whether analog or digital computing was best, he responded—"Get them both. You need all the help you can get." If I were asked whether field theory or circuit theory is best I would have to answer a little differently. You need to use them both. One alone is not sufficient.

Our group of designers were moved to Transformer Engineers, another Pasadena company. I invented a way to build dc amplifiers and avoid voltage regulation using two blocks of ac gain, feedback, and a mechanical chopper to reduce the size of a channel but management did not want to enter this market. Four of us including George, left and formed our own company called Dynamics Instrumentation. The new design was a viable product and within a year I had invented a reasonably good differential amplifier that provided a bandwidth of 10 kHz. I began to understand that each amplifier channel needed its own power supply or there would be system oscillations. Following soon were the days of the first transistors and the instruments were getting smaller and more sophisticated. This was the period when rocket engines were being tested for space exploration. As semiconductors developed I was eventually able to build differential amplifiers without mechanical choppers that were extremely stable.

I began to write small articles that were used as sales tools. There was essentially no written material available for engineers to study and these articles were very effective sales tools. These articles about shields, the shielding of instruments and the grounding of cable shields, led me to write my first book. Much to my surprise Wiley accepted my manuscript for publication. What I did not know was that the engineers buying the instrumentation had little to say about how the facility would handle the cables and their terminations.

Of course, grounding and shielding involves far more than testing aircraft or rockets. It involves electronics in medicine, research, manufacturing, and computers. Instruments had to function in adverse environments and survive when there was nearby lightning activity. I learned how to protect input circuits from significant overloads. Input noise levels of 2 microvolts rms in 100 kHz were accepted standards and yet the instruments were protected even if the power line were connected to the input. Instruments could operate with 300 volts ground potential differences between input and output.

As a principal in my own company I had the opportunity to visit and talk to engineers in facilities all over the United States. They were very willing to show me their testing installations and rooms full of electronics. A typical rack would be opened up and all I could see was a bundle of cables a foot in diameter coming out of the floor and fanning out to hardware. It was impressive and I had no idea of what this meant from the standpoint of performance. All I knew was that if I tested an instrument on my bench it would meet my published specifications. All of this contact gave me an understanding of electronics which I did not get in college.

Dynamics came to an abrupt halt as a company in one of the big cutbacks in defense spending. I went to work for a company that supplied equipment to the telephone industry. I remember being shown the special earth ground that had been brought up into the lab. This was apparently standard practice in the telephone industry. The chief engineer felt strongly that this connection would ensure that electronics could be tested in a "quiet" environment because this ground was available. I did not respond because all of my experience indicated that this ground was of little importance. To me it was just another earthed conductor along with water pipes, gas lines, and electrical conduit. The only difference was they each had a special title.

The telephone system in those days rang bells by placing a voltage between one telephone line and earth. The telephone systems would work even though there was a power outage as the system ran on batteries where the positive lead was grounded. The ground conductor I was shown was a part of telephone practice. I later became chief engineer of that company and never once did I use that special ground.

I understood the philosophy of grounding analog amplifiers. I had built analog circuits and found that "hum" pickup was reduced if I grounded my circuits to a nearby conduit. My approach in instrumentation had been that if I had to use special grounds to get performance, my design was inadequate. What I was learning was that this search for "good ground" had caught on and was accepted as good engineering. I found the idea that a good ground would work for a facility regardless of its size ridiculous. But as I traveled the country I found this idea had taken hold and was accepted as good practice.

As an April fool's joke I submitted a news article to an electronics magazine describing that the perfect ground had been discovered near Nome Alaska and it was being protected against contamination by a fence.

I had occasion to visit a deep space antenna facility in the Mohave desert. There was a circle of buildings with a central point acting as ground for all the electronics. This ground was made from buried copper rod. The power for each building was supplied through a nearby distribution transformer that was grounded (earthed) at the transformer per the electrical code. The problem occurred during thunder storm activity. Lightning would strike the earth causing a potential difference between the central ground and the utility ground at each building. A nearby lightning strike would place thousands of volts across the coils of the distribution transformer. The result was that these transformers were being blown up. This grounding practice was written into law and quality control would not accept any change to this practice. What bothered me the most was that the engineers we talked to had no idea of what was wrong. They were interested in the performance of individual pieces of hardware and the facility issues were not their domain.

This single-point grounding practice was what I saw in most major defense installations. All the shields for all the signals were bundled together and connected to this "good" ground. The simple physics in my book showed why this could not work, but no one paid attention to what I considered to be obvious. Buildings are not circuits and the single point grounds that worked in many applications could not function in these large environments. They thought they knew where the interfering current entered the sink but no one asked where the current came back out of the earth. The idea of skin effect was never considered.

The first electronic instruments I designed were full of shielded conductors. This is labor intensive and adds a lot of bulk and cost to a design. My last designs used no shielded conductors. In feedback structures, most of the interference is cancelled. Knowing how to route signals and how to control impedance levels, use differential techniques made for simple layouts. In the end, there were no calculations—just technique and it worked.

Making things work well and at the same time keeping things simple is a challenge. There are only so many coupling mechanisms. Understanding that both interference and signals obey the same laws is key to finding answers. When I viewed a facility problem my circuit experience served me well. I could see the problem in terms of the spaces not the conductors. The rules people usually follow involve conductors without regard to the geometry. For example, the service entrance to a residence requires a grounding conductor. The intent is to provide a path for lightning to earth. If the conductor is routed around a door or along a wall for any distance to get to a water pipe, it will not function even though it passes inspection. People follow rules. They have no choice.

I visited a facility in New Mexico that assembled live missiles for the air force. The fear was that an ESD event might arm a missile. I was called in because a missile had somehow been armed. Every worker at a bench wore a conducting wrist band that was connected to his or her bench and seat. The floors were conductive and the walls of the building were corrugated steel. This facility was located in an area where lightning frequency is high. When I arrived, there was a group of inspectors in white smocks and with clipboards touring the facility looking for any rule violations. I was told that if there was a chance of lightning, the building would be evacuated.

I noticed that the building had lightning arrestors on the roof eaves and that there were down-conductors from these arrestors to grounding posts. Then I saw a television antenna that was taller than the lightning protection. What I saw next was that lightning would use the corrugated steel walls as a down conductor, not the down-conductors provided. The problem then was that lightning using this path would arc to earth at floor level. Missiles were located on the other side of the wall just feets from the points of possible arcing. I recommended that the building steel be grounded at points around the perimeter of the building. Any arcing at this level could easily arm a missile. The lesson here is that following written rules does not always solve problems. Common sense is important.

My music provided another side to my life that was unusual. I play chamber music and this has brought me in touch with some very interesting people. In Pasadena, I played music with Rupert Pole who was married to Anais Nin, the famous diarist. I played music with Alice Leighton, wife of Robert Leighton, who helped write the famous Physics Lecture Series with Dr. Feynman. Stuart Canin, who became concertmaster of the San Francisco Symphony was concertmaster of the GI Symphony and played for Roosevelt, Churchill, and Stalin in Potsdam during our stay in Berlin. At the time, this was done in secret.

I had a similar experience when the GI orchestra was stationed in Frankfurt. I played quartets (background music) for a group of officers holding a fancy dinner party at a castle on the banks of the Rhine river. They were living it up with a group of army nurses. I would have never seen the inside of a castle if it were not for music.

My first book was published by John Wiley and Sons in 1967. This was a period in time when rocket engines were being developed. The title of this book was *Grounding and Shielding in Instrumentation.* The main problem I addressed was related to limiting interference to microvolts when the input signal lines ran thousands of feet between rocket stands and instrumentation bunkers. Designing instrumentation amplifiers using vacuum tubes that worked in this environment meant I had to understand interference from both internal and external sources. I had solved the instrumentation problems but the users had to understand their role in handling long input cables. It is a long story but in later years I found out that my recommendations were largely ignored. Performance was impacted but obviously systems worked well enough to get men on the moon. I have rewritten *Grounding and Shielding* every 10 years and the 6th edition just came out in 2016. I have no doubt that many of the shielding issues that I discuss in my books are still not being handled correctly. What often happens is that as new technology enters the scene the old problems simply fade away. It is not comforting to realize that progress often takes this form. A lack of understanding persists and it will cause problems in some future effort. We seldom argue with a practice if it works. I do not like invalid arguments that supported bad practice. It leads to problems in the future.

I found it very rewarding to put my understanding into the written word. Writing it down on paper forces me to be accurate. My understanding in many areas was challenged and I had many things to learn. My first book brought me consulting jobs as well as opportunities to teach seminars. The more contact I had with engineers the more convinced I became that they were not using the physics they had been taught in school. It was through my books that Dan Beeker found me and asked me to do a few seminars for his company.

These early seminars were not directed at the world of circuit boards. They covered areas of power distribution, radiation, and instrumentation. The subjects of grounding and shielding were treated for both analog and rf circuits. Facility design, lightning protection, ESD, screen room design, and radiation from antennas were all key topics. My talks had one thing in common: the problems of interference were all field related. For Dan's audience, I began to include topics related to transmission lines on circuit boards. Dan heard my talk dozens of times. I contradicted many of his pet ideas. I told a story that was different than other speakers. Much to his credit he began to see a light and he began changing his approach to board layout. He has become very successful at trouble shooting problems using a new set of ideas. His boards work the first time.

Electronics has changed dramatically over the last 50 years. My early experiences were analog. The world is now largely digital with the communication between devices being wireless. The thing that has not changed is that circuit board layouts can have problems. The signals have greater bandwidth but the same physics of electricity and magnetism applies. This book is intended to provide tools to designers so that logic circuit boards can function at maximum clock rates and so they will function reliably and pass radiation tests. It is also intended to give IC manufacturers a system view of logic operations.

There are several levels to the design of logic circuit boards. The design involves selecting processors, memories, clock systems, power supplies, buffers, and drivers. A second aspect is the layout. This includes the number of board layers, the location of components, type of board materials, trace characteristics, the number of ground planes, and how power is distributed. A complete design requires interconnecting the pins and pads, using conducting planes, traces, and vias. This set of connections seems like an elementary problem, not one requiring engineering attention. There is a lot to "wiring" a circuit board. It is not only a problem of interconnecting logic but also one of moving energy.

What makes the problem difficult is that we are getting to the point where every picosecond is important. This is no longer the domain of circuit theory.

I could provide the reader with a set of rules to follow and then there would be no book to write. I feel that it is necessary to explain electrical activity in terms of electromagnetic fields. This way the reader will have two sets of tools to use in the future. I want to do this so that no matter what the future holds the reader will be prepared. I want to present this material in a straightforward way. There is no way to avoid some mathematics. The one thing I know for certain is that fields explain things that circuit theory misses.

Fields are not components and they cannot be seen. Progress is usually made by using old materials in new ways. The changes that are taking place are adding to logic rates and radiation, and they are in the direction to reduce signal integrity. Progress is made by keeping the fundamentals in full view. The fundamentals I am referring to are the electromagnetic fields that define all electrical phenomena. We are designing products where new tools must be put in the tool kit. My goal is to point out these tools. You have to learn how to use them. There is a recognition that a new technology is needed to make further progress. Until that happens we need to make optimum use of the products we have. Whatever is invented, it will be based on physics and circuit theory will have to step back a little further.

A better understanding of the electrical world did not come until I started to design hardware and tried to reduce noise levels, increase bandwidth, and interface with utility power. At every twist and turn there was some parasitic effect that could only be explained by using my physics background. It took time but I learned that every component had both an ideal and a practical character. Voltage sources were not zero impedance and switches were complex objects. I soon learned to be suspicious of everything including conductor geometry. Every circuit was unstable until I ran tests. I designed some logic circuits and observed that there were many new performance limits to consider. I watched wire-wrap technology evolve and disappear and then multilayer boards take over. As the technology evolved I was able to present my understanding every 10 years in 6 editions to my *Grounding and Shielding* book. In my seminars I started to slant materials toward circuit boards that perform logic functions. This led me to spend a lot of time working on transmission lines that handle digital information. Most of the material that exists in the literature relates to sine waves and little is written as it relates to step functions. I had some questions about Poynting's vector and step functions and got no help from my physics friends. Eventually I put together a picture of what happens when waves reflect at a discontinuity. It took time but I finally realized that what was happening on a logic board was the movement, storage, and conversion of energy. I finally understood that if you give nature an opportunity to move energy, you can meet goals in moving information. This energy is moved in fields in the spaces between conductors. The traces act as guides that control where the energy can flow. This book is written to describe this energy flow in detail. It may surprise you that there is so much to say.

The path I have taken to understand electrical behavior is perhaps unique. There are few books devoted to the relationship between electrical interference and the basic physics of electricity. I know the pressure to understand the world is there because this is what engineers do. They make things work when there are forces at play that limit performance. I am familiar with the trial-and-error approach and I also know that explanations of why something works are often in error. I know because I have been a part of this process. Fortunately, I had a good education and I keep challenging my own explanations. The truth is that in the process of learning I have often been in error. My excuse is that essentially no one has offered me criticisms or advice. The physicists I know have never built their own circuit boards or wound their own transformers, so they have had no experience solving the types of problems I was forced to deal with. I was very much on my own. There was always a feeling of recognition for my writing efforts but there has been essentially no feedback. One of the biggest problems we all have is discussing our inadequacies. Another problem is taking positions that are contrary to accepted practice. It is always easy to use the local jargon and accept the opinion of the boss. The lore that prevails is working (sort of), so why be contrary. I also know that spending a billion dollars on an approach does not make it right. It does make it very difficult to criticize.

I feel I have a good picture of interference processes. When I attend a conference, I can usually tell when a speaker has a convoluted understanding. The paper is published, the audience applauds, and life goes on. If I give a presentation, the listeners react the same way. Because there is a big gap in understanding, both presentations are probably ignored and the listeners are left looking for understanding. They will come back next year hoping for a better experience. My hope is that I can open a door just far enough so that a correct understanding can take root.

Emotions often run high in areas where opinions differ. I understand the effort that it takes to be an engineer. It is human nature to defend a position that has taken years to establish. There are engineers that feel strongly that field theory explanations are the wrong approach. I feel just as strongly that they are the ones in error. Many are successful designers that are not afraid to speak out. My experiences tell me to remain firm. Nature is not going to change one iota just because we explain phenomena in different ways. Many designers invent new languages that seems to catch on like a virus. This practice unfortunately places impediments in the communications channel. I understand the dynamic nature of language but I also realize the need for accuracy and stability. No one is changing the words voltage or inductance, but often engineers forget that these words are defined in terms of field theory.

Learning electricity is an ongoing process for engineers. We cannot see electricity, so we must live with a stream of interconnected concepts. Sometimes, what we think we understand is flawed. Until we are challenged, these flaws are not recognized. The first thing we try to do is fit the facts we see to the ideas we feel comfortable with. I remember in college studying the same material over and over, each time adding a level of complexity. The same thing happened when I went to work designing hardware. Some of the learning involved applying new materials and some of it was related to a better understanding of the past. Lately, learning has involved going back to basic physics as circuit theory is showing its limitations. Going back is only a part of the process. In my early attempts at design I had little idea of just how to proceed. First inclinations were to experiment with the materials at hand. When experimentation was not practical, I asked my boss.

I remember the first time I encountered a shield in a power transformer. I was not sure where to connect this shield. There were several choices including the instrument enclosure, common signal, or equipment ground. I asked by boss who was a very respected engineer. His response surprised me. He proceeded to light up his pipe. The delaying tactic said he did not know. Later in my career I realized there was no simple answer. We were building high-quality equipment and a shielded transformer represented this higher quality. In fact, it made little difference where this one shield was connected. I now realize he was being honest and that we both had a lot to learn. It took years before I found the answer to my question.

Later in my career I worked on differential amplifiers that required the use of transformers with three shields. I built my own shielded transformers and found out how to use this shielding to build a wideband differential instrumentation amplifier. I wrote the specifications and got a transformer house to make the transformers. The company saw this as a business opportunity and added "Isolation Transformers" to their product line. These transformers found application in the early computer installations that were being used in the defense industry. I wrote about these transformers in my Grounding and Shielding books.

One day I was scanning an electronics magazine and noticed an ad by this same company for a four-shield "Isolation" transformer. I had no idea of how four shields might work so I called the president of this company hoping to learn something about shielding. I asked the obvious question "Where do you connect this fourth shield?" The answer was honest. "I do not know." Then I asked the question "Why do you offer this shield?" The answer was equally honest. "It sells more transformers." That answer has stayed with me for 50 years. Looking back, I can see that there is an element of desperation in engineering that requires action even if the reasoning is not well founded.

It became obvious to me that a building was not a circuit. There was no way to use my meters and an oscilloscope to measure its properties. Even if I could describe the building as a circuit, there was no way I was not going to redesign it. A building and its conductors are a "given" and I was not going to change buildings. Experimenting with buildings is not an easy thing to do even for governments. I decided I had to design "islands" of space that I could control. Meanwhile, I saw efforts to design buildings that were supposedly "quiet," "clean," or "radiation proof" that no one could test. As we all know, engineering cannot be based on hope or unfounded ideas. Testing buildings is not an easy thing to do.

I have reached the conclusion that there are significant problems in observing electrical behavior whether it be a circuit board or a building. There is much that cannot be seen. Of course, if the circuit works, we are not apt to be too critical. We cannot see under traces or on traces between board layers. We cannot see inside of components or inside of conductors. We cannot see current distributions, let alone current flow. These limitations are obvious. What I have just mentioned only touches the surface. What we cannot see and what we do not look for are the shapes of the fields that occupy the spaces between conductors. These fields carry the energy yet in more ways than one they are inaccessible. Voltages give an indication of electric field, but we are blind to the magnetic field that must be present. Some of this blindness is of our own doing. It is possible to infer what is happening but it takes effort.

It is possible to study the physics of electricity and not see its relevance. The mathematics is one thing but sensing the movement of energy is at another level of understanding. The energy stored in the earth's gravitational field is enormous. We walk around in this energy field and pay little heed until there is an event. The fact that this field warps time is completely out of our picture. All of this is at the heart of nature and it takes a real effort to recognize that we live in fields our entire life. To put it simply, we are blind to fields unless we make a very special effort. Fields dominate the first chapter. These electric fields are the very fields of life as they form molecules from atoms. These same fields operate our circuits and we know a great deal about them. That does not mean it is impossible. We came into this world not knowing the shape or size of anything. It all had to be learned. Electricity is no different. The big problem is we cannot touch or feel it. This means we must create our own images. Often these images need correction and that takes courage and time. Habit is at the heart of our survival. Not all habits are in our best interest. They are hard to change.

The beauty of physics is that it explains the present and shows us the way to the future. It is not dedicated to any one discipline. Physics is constantly being challenged and matched with experience. The links between experience and what is basic is not always obvious. It takes time and a strong desire for understanding before the cobwebs are removed. This effort is very personal in character and usually does not transfer to others working in the same discipline. When the understanding solves problems, there is often a financial benefit and that does get attention.

This book presents some ideas that I have not seen published or discussed. It is the result of trying to connect together all the pieces of my understanding. In one sense it says I am making progress. In another sense it says I have a lot more to learn. What I have learned I have tried to put into words. It is not easy to accept new ideas and connect them with past experiences. It takes effort and time. I hope this book can help you understand better the electrical world. May your designs work well the first time.

Ralph Morrison
San Bruno, CA—2017

1

Electric and Magnetic Fields

Buildings have walls and halls.
People travel in the halls, not the walls.
Circuits have traces and spaces.
Energy travels in the spaces, not the traces.
Ralph Morrison

Fast Circuit Boards: Energy Management, First Edition. Ralph Morrison.

1.1 Introduction

I have written many books that discuss electricity. The sixth edition to my *Grounding and Shielding* was published in 2016. Each time I start to write, it is because I have added to my understanding and I feel I can help others solve practical problems. The digital age is upon us and it is making demands on design that are not covered by circuit theory. In one sense, logic is easier to handle than sine waves as there is no phase shift to consider. As clock rates rise there are a host of other problems that must be considered. The tools of circuit theory are in many ways a mismatch for fast circuits and some new methods need to be introduced. Circuit theory does not consider time of transit and this is a key issue in logic design.

The insight that has led me to write this book involves one key word and that word is *energy*. Engineers are taught that stored energy can do work and that there is conservation of energy. What I want to point out is that **all** electrical phenomena is the movement or conversion of energy. When a logic signal is placed on a trace, the voltage that is present means that energy was placed on a capacitance. This energy had to be moved into position and it had to be done in a hurry. It cannot be done in zero time as this would take infinite power. The movement of voltage in circuit theory does not mention moving energy and that one fact leads to many difficulties. For example, energy is not carried or stored in conductors. It is carried and stored in spaces. This requires a different point of view. Providing this insight is the reason for writing this book.

I am writing this book to help you avoid problems when laying out digital circuit boards. This first chapter is about the fundamentals of electricity. You may opt to skim over this material but I would hope to change your mind. I want to show you how electricity works but not from a circuit theory viewpoint. I want to show you how to use the spaces that carry energy between components. This may be a new way for you to view electricity. Circuit theory pays little attention to spaces. Conductors outline the spaces used by nature to move field energy.

I am writing this book to describe the role fields play in moving energy on transmission lines. Most of the literature relating to transmission lines revolves around sine waves. There are insights to be gained when step functions are applied to moving energy. These insights can help you in today's designs and prepare you for the future which is headed in the direction to move more and more digital data. Whatever the future holds, fundamentals apply.

This opening chapter will help to set the stage for discussing transmission lines. Once you see how nature really works, you will find it easy to accept the material in the later chapters. I have spent my career learning and applying this material and I know why basics come first. What is even more important is connecting these basics to real issues and that is not always easy. That is my objective in writing this book.

It is worth a moment to describe a problem in understanding. In reading posted online material, the language that is used often implies a picture of electrical behavior that is invalid. A writer may say that "—the return path impedance is too high." The impedance of a transmission path is really the ratio of electric and magnetic fields involving the space between the forward and return current conductors. It is incorrect to treat the forward and return paths this way. It takes care in the use of language and a clear understanding of basics to communicate clearly. In this book I have chosen my words carefully and I hope they are read with the same care. I realize there are strange explanations that must be unlearned and that is not always easy to do. Currents do not travel one way on the surface of a conductor and return down the middle of that same conductor. I will not spend time trying to argue this point.

There are many ways to learn electronics. Diagrams that include connections between components are common. These diagrams make no attempt to suggest that energy is flowing in the spaces between the connections. It is through experience that an engineer learns that component locations and wiring methods are important. Understanding the mechanisms involved is presented in this book. What is hard to accept is that there are a lot of habit patterns that develop over the years. Some of these patterns may work at the time but lead to problems as the logic speeds increase. Looking back, I can say that many published application notes I tried to apply were invalid.

A good example of the problem exists in the wiring to a metal case power transistor. A metal case provides for heat transfer as well as a connection to the die. The fields that move electrical energy cannot enter or leave through the metal case. Fields can only use the space around the connecting pins. This results in fields sharing the same space which is feedback. I remember discussions on how to mount these devices that involved surface roughness but never a discussion of field control. Because the circuits worked, I did not question the printed page.

If surface roughness was a consideration, it also meant that the characteristic impedance of the energy path was important and this was not mentioned. This issue will become clearer in later chapters.

We live in an electrical world. Atoms are made of electrical particles. All the energy that comes from the sun is electrical. As humans, our brains and nerves are electrical. We have used our understanding of electricity to communicate, travel, and entertain ourselves. We have learned to use electricity to light our cities and to run our industries. Today's electronics gives us computing power that invades every aspect of life.

INSIGHT

Long before the first light bulb was invented, physicists were working on understanding electricity. In 1856, James Clerk Maxwell presented to the world a set of equations that described the physics of electricity. This is recognized as one of the greatest achievements in the history of science. Even with all the computing power that is available today, these equations are difficult to apply to many practical problems. These equations describe electrical activity in terms of fields, not circuit elements. Remember that circuit elements as we know them had not been invented when these equations were developed. These equations covered electromagnetic radiation long before there was such a thing as radio. Another remarkable fact is that these equations also agree with Einstein's relativity concepts.

Most of the energy we get from the sun each day is electrical. We also get energy from the hot core of the earth and some of it comes from the gravitational field of the earth/moon. The sun has stored energy for us in the form of coal and petroleum. On a daily basis, the sun's energy grows our food, heats our land, evaporates ocean water, and moves our atmosphere allowing us to live our lives. The actions of nature are in effect unidirectional. Sunlight is converted to heat but not the other way around. Water runs downhill except where trapped in pools. Air runs out of a balloon. We know how to put air back into the balloon, but it requires some other device that releases stored energy.

The game of life is the motion of energy through a vast complex of matter. One of our engineering goals is to perform logic functions. We can accomplish this task by providing pathways for nature to move energy. This energy flow is in the form of electric and magnetic fields. We use these fields because they are fast, efficient, inexpensive, repeatable, and available. The problem we face is how to do this with the materials we have at hand. To be effective, we need to know the rules nature will follow as we select and configure materials to build our circuits. Nature is very consistent.

A good place to start is to move some energy. A dropped stone gives motion to air as it gains kinetic energy. On impact, some of the energy is converted to sound, mechanical vibration, and heat. The total energy is conserved but it is no longer energy of position in a field of gravity. The energy cannot return to its initial state. In the case of a pendulum, potential and kinetic energy can move back and forth, but there is always friction that will remove energy and in time the pendulum will stop moving. I like to call this nature running down hill. The basic explanation involves a concept known as entropy. All systems tend to "spread out" if given an opportunity. We rely on nature to take this action. When we connect a resistor to a battery we expect it to get hot. If we poke a hole in a pail of water, we expect water to drain out. If we connect a voltage to a pair of conductors, we expect energy will flow between the conductors. Nature does this the same way over and over, never changing the rules. The way energy moves on a circuit board is the subject of this book. I want to show you nature's rules so we can perform needed tasks. It sounds trite to say this but it is important. We can follow nature's rules but she will not follow ours. I do not care who signs the documents.

Circuit theory is key in the education of electrical engineers. The physics of electricity is taught but the connection to practical issues is not always a goal. This is not surprising as there are so many possible specialties and applications. The basics are taught and it is assumed the engineer will make the needed connections on his own. Electrical engineers are taught circuit theory that handles linear and active circuits using sine waves up to frequencies of several MHz. The basic assumption made by circuit theory is that an analysis based on how ideal circuit elements are interconnected tells the full story. This viewpoint sidesteps the issue of field theory and energy transfer which are needed to move logic. At frequencies above a few MHz, circuit theory begins to show its inadequacies. To add to the problem, logic signals are not sine waves which are the basis of circuit theory. Take away circuit theory and stay away from field theory and it becomes a big guessing game. Engineering and guessing are not compatible. Engineers that start their design activities using circuit theory rarely try to use field theory to solve problems. A change in approach will only occur if someone understands the need for change.

The fundamentals of electricity are properly expressed in terms of partial differential equations. These equations are difficult to use even with computer assistance. The solutions that are provided revolve around sine waves and logic has voltage signals that are step and square waves. Even if the engineer is a skilled mathematician it is not practical to find exact solutions using field equations. The mathematics tells a story that is very compact. Very few can understand the full story by just reading equations. Until recently, a common-sense approach to interconnecting components on a circuit board was quite satisfactory. This approach has limitations specifically at high bit rates. When there are problems, the only approach is to go back to basics. It turns out a lot of insight can be gained by applying the basic principles of physics. The simplifications offered by circuit theory have reached a limit. What is not appreciated is that circuit theory lets us make many false assumptions that become habit patterns. The difficult part of applying physics to fast logic is not the new ideas, but the letting go of assumptions that have been made and that we are not aware of. The most troublesome assumption is that conductors carry energy. Conductors direct fields and fields carry energy in spaces. I will say this over and over again. This is the inverse of what most people sense.

When I started to design circuits, I soon learned there were issues that had to be faced at every corner of my reactance chart. Small parasitic capacitances were everywhere. Resistors above 100 megohms were not resistors above 100 kHz. Magnetic materials were extremely nonlinear. I found out that capacitors had natural frequencies and that any feedback circuit I built was guaranteed to oscillate. I found out that power transformers coupled me to the power lines and test equipment connected my circuits to earth. Every time I started working on a new product the real world presented me with a new set of problems. Most of these problems were not discussed in any textbook. These were problems that were not helped by using mathematics. Laying out digital circuit boards is no different. There is a lot to learn before designs are free of problems. This is particularly true as clock rates enter the GHz range. Understanding compromise requires understanding fundamentals. Good engineering requires a lot of compromise. The problem is compromise must be based on understanding.

Logic circuits are mainly the interconnection between various integrated circuits. It takes experience to realize that these interconnections must be designed. Connecting logic components together is the subject of this book. Until problems start to appear, the designer is not apt to think of applying any change in approach. In analog design, I found out that microvolts of coupling occurred if I was not careful in interconnecting components. Usually this meant avoiding placing any open spaces in the signal path. A similar set of problems occurs in logic. The voltage levels that cause errors are greater but so is the frequency content. In both cases, it is the same basic electrical phenomena that must be understood. In very simple terms, keep all transmission fields under tight control.

In the digital world, it is not practical to experiment with layout. A redesign is expensive and it takes time. Tests to prove conformance with regulations are expensive. Making measurements that can isolate problems also has its difficulty. A practical sensor can pick up signals from a dozen sources. An identification of which source is causing a problem is not usually possible. Wiring on the inner layers of a circuit board is not accessible. Parts of the transmission are inside an IC package and there is no way to view these signals. The list of difficulties is long. All factors need to be considered at the time the traces are located. These factors are the subject of this book.

In the following chapters, I discuss logic transmission in terms of step functions. There is no simple mathematics that fits this type of signal. Circuit theory and sine waves were made for each other but logic requires different tools. It is not all that complicated but it is different. It takes time to study, review, and accept a new set of ideas and routines and put them to practice. What may not be appreciated is that there is insight that can be gained by using step functions that is hidden when sine waves are used. I point out these insights as I write.

We use some mathematics to discuss the principles of field theory. This will help in explaining what is happening on a logic circuit board as the clock rates cross into the GHz region. An understanding of differential equations and vector analysis is not a requirement for understanding the material in this book. I use both language and mathematics to explain each topic. For some, the language of physics will be new. Both circuit components and the way they are interconnected will be important. An understanding will eventually lead the designer to follow good practices. Taking this view will make it possible to design circuit boards without making false assumptions. Once the rules are known, they become the new habit pattern. The designer must keep the basics in mind.

I assume the reader is familiar with the language and the diagrams of circuit theory. I want to caution the reader that diagrams often suggest meaning that is misleading. A good example is the symbol for an inductor. A coil of wire has inductance but so does a straight conductor. A trace over a ground plane also has inductance as well as a volume of space. Extending the understanding of inductance is an objective in this book.

The symbol for a capacitor is another example. The symbol with its symmetrical connections gives the impression that current arrives at the center and spreads out symmetrically. What really happens is that current enters at one end and is associated with a magnetic field. As you will see, a capacitor looks more like transmission line than a component. This is a picture not suggested by the symbol.

Another example of how we are misled is the diagram for a transformer. The last turns of a primary coil are capacitively coupled to the first turns of a secondary coil. The diagram implies some sort of symmetry which is rarely present. I am not suggesting a different symbol only an awareness that we tend to make assumptions. When we read we do not check spelling but a misspelled word sticks out. This same understanding can take place in board layout. Schematics and wiring lists tend to hide issues of layout. It may become necessary to add a requirement that a signal flow chart be a part of each design. It is an important part of a design.

Logic has become so complex and proprietary that the structure inside a component is rarely made available. The compromises taken by the manufacturer are not discussed with the user. The user assumes that the logic will function at the specified clock rates. The manufacturer decides just how much information the user needs to know.

The language of electricity is not always clear. There are words we use with literally hundreds of meanings. Two examples are the words ground and shield. Some words can represent a vector or a scalar and without help, the reader is left guessing. Both the writer and the reader have a responsibility to be careful in treating the written word. Often the meaning depends on context. Read with an open mind. Hopefully the picture will grow in clarity as more and more examples of the issues are presented. The glossary at the end of each chapter is there as a reminder of how much attention we must pay to language.

In any rapidly changing field there is a tendency to invent language, whether needed or not. Acronyms often become new words. In this book, an effort will be made to use well-established language including terms used in physics. Words change their meaning through usage and sometimes the meaning is transitory. I want to use language that makes sense to old-timers as well as to the next generation of engineers. I expect that newcomers will need all the help they can get. I will avoid acronyms as they are not always stable. I have attended meetings where disagreements have occurred because two definitions of one acronym were involved.

This book does not discuss logic diagrams, the choice of components, software methods, or clock rates. Our interest is in describing how to move energy so that the functions taking place are performed without introducing errors or causing radiation.

INSIGHT

I mention in Chapter 2 that forward energy flow and reflected waves travel in opposite directions. The fact that they are both traveling at the speed of light means that relativistic effects are involved. This may be the next area of physics that must be considered as we move into the GHz world. It is interesting to note that our GPS systems must be corrected for the relativistic effect gravity has on electromagnetic waves. Without this correction, the system could not work. At this point it is important to recognize that change is on the horizon. I want to help make the transition wherever it takes us.

1.2 Electrons and the Force Field

An electron is a basic unit of negative charge. It has the smallest mass of any atomic particle. In an atom, electrons occupy quantum states around a nucleus. Chemistry is a subject that treats the formation of stable molecules when elements share electron space. The simplest atom is hydrogen which has one electron and one proton. All other elements have both protons and neutrons in a nucleus and electrons in positions around this core. A proton has a positive charge that exactly balances the negative charge of an electron. When an electron leaves an atom, the empty space behaves as a positive charge. In effect the field of a proton has no place to terminate inside the atom. Another way to treat positive charge motion is to compare it to the motion of an empty seat in an auditorium. If a person moves left into an empty seat the space moves right. In a semiconductor, the absence of an electron is a hole that has a different mobility than an electron. In a conductor, the mobility of a vacant space is the same as an electron. In electronics, we consider that all protons are captive and will not leave an atom to partake in current flow. We also assume that the atoms in a conductor do not move any distance.

Copper atoms have 29 electrons with 1 outer electron that is free to move between atoms. Avogadro's number tells us that in 63.54 g of copper there are 6.02×10^{23} atoms. This is the number of electrons in this amount of copper that can move freely between atoms. I am giving you these numbers so that you can better appreciate what really happens in a typical circuit. I think you are going to be very surprised by the story I am going to tell you.

We know there are forces between electrons. When hair is combed on a dry day, the individual strands of hair will separate and stand straight out. In a clothes drier, the rubbing action can move electrons between items of clothing that can create a glow seen in a dark room. We have all seen the power of lightning[1] where rain strips electrons off of air molecules and carries charge to earth. In the laboratory on a dry day, it is possible to rub a silk cloth on a glass rod and transfer electrons from the rod to the silk. The rod is said to be charged positive. Touching this rod to hanging pith balls will cause the balls to swing apart. If the cloth rubs a hard rubber rod, then electrons are added to the rod. When the hard rubber rod touches one of the pith balls, the two balls will attract each other. If the pith balls are metalized, then the charges will move on the conducting surface until there is a balance of electrical and mechanical forces. In this case, the electrons cannot leap out into space and the forces appear to be between the pith balls. These forces are actually between charges. These forces are shown in Figure 1.1.

When there are forces at a distance it is usual to call the force pattern in space a field. In the example above, the force field is called an electric or E field. Fields have both intensity and direction at every point in space. The field we are most familiar with is gravity. On earth the gravitational field pulls each of us toward the center of the earth with a force equal to our mass. 4000 miles out in space, the force would be reduced by a factor of 4. The electric field behaves the same way. A negative charge located on a conducting sphere of radius R creates a

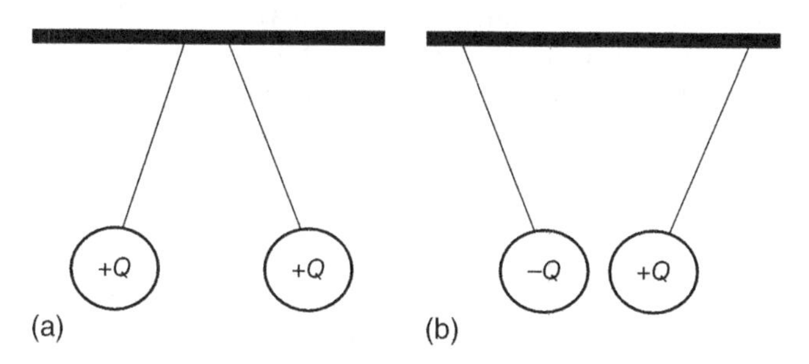

Figure 1.1 The forces between charges. (a) Repelling force and (b) attracting force.

1 Radiation from the sun strips electrons from gas molecules in the ionosphere. The potential from this surface to earth is about 400,000 volts. This surface provides a current flow to earth and a drain into space for accumulated charge. The rain that strips electrons from air causes a voltage gradient near the earth that is approximately 100 volts per meter. This voltage varies during the day as the result of lightning activity which is maximum in Africa at 4 p.m. Greenwich mean time. There is a steady flow of current to earth as a result of this gradient. It is estimated that this current world wide averages about 1800 A. There is a steady flow to this current as well as bursts from lightning.

force field. A small positive test charge near the sphere will be attracted with a force F. If the test charge is moved out to a distance $2R$, the force will be reduced by a factor of 4. We will have a lot to say about electric fields. Later we discuss the magnetic field, which is a force field of equal importance. This field is quite different because the motion of charges creates forces at a distance.

INSIGHT

To put the electric force field into perspective, I want to refer the reader to an observation made by Dr. Richard Feynman in his book Lectures on Physics. He asks the question: How large would the force be if we were standing 2 feet apart and 1% of my electrons were to repel 1% of your electrons? Would it knock you down? Much more. Would it lift a house? Much more. Would it lift a mountain? Much more. The answer is astounding. It would lift the earth out of orbit. No wonder the electric field force is labeled a strong force.

INSIGHT

When we observe a typical circuit, nothing moves. There are no discernable forces. That means that hardly any of the electrons that are free to move are involved in electrical activity. The number of active electrons can be compared to a cup of sand in a mile of beach. It is an amazingly small percentage. Our sensibilities are challenged because for all practical purposes the charges that move in a typical circuit move in a continuous or fluid manner. This is the reason I brought up Avogadro's number. **In a typical trace there can be a trillion electrons in motion, yet the fraction of those in motion is only one part in a trillion.**

INSIGHT

Electron flow is sparse and yet continuous at the same time. Thinking in terms of these large numbers is not easy. Viewed from one free electron, the nearest free electron is far away and hard to find. Viewed from our perspective the number of free electrons in motion appears continuous.

To put a trillion into perspective, there are 7.35 billion people on earth.

Here is another observation that is worth discussing. The chemical properties of elements depend on the number of electrons available to interact with other elements. In our circuits, there is no evidence that the properties of materials change when large currents flow. We do see property changes when materials are dissolved in a solution. That is because all of the atoms are involved, not a very small percentage.

Now you know that the percentage of electrons involved in circuit activity is very near zero. It is easy to show that the velocity of electrons in a typical circuit is under centimeters per second. This is technically known as electron drift. We know that signals on a circuit board travel at the speed of light. This says that we should focus on the field that surrounds the electron, not on the electron itself. This is because electrons are a slow mover in every sense. The problem is they keep bumping into atoms. The force field is potent as we have discussed.

The changing of an electric field creates a magnetic field. This fact is stated in Maxwell's field equations. The fact that the electric and magnetic field act together to move energy is at the center of all electrical activity. We discuss the magnetic field after we have discussed the electric field. Then we discuss how the two fields act together to move energy along transmission lines. Of course fields also move energy in free space but not at dc.

On earth, we live in a force field called gravity. The force field around a mass is proportional to that mass and falls of as the square of distance. In gravity, the fields between masses always attract; they never repel. A gravitational field is called a weak field. After all it takes the mass of the entire earth to attract a 150-pound person with a force of 150 pounds. There are several other force fields in nature and they exist in the nucleus of the atom. These fields are a topic for a course in nuclear physics.

INSIGHT

Orders of magnitude are often ignored as issues by engineers. A man can run at 4 mph. An auto can go 40 mph. A jet can go at 400 mph and a satellite goes at 4000 mph. That is a range of three orders of magnitude. Going from 1 MHz to 1 GHz is also three orders of magnitude. Avogadro's number is 23 orders of magnitude. It is hard to grasp the meaning of such a large number, yet that is the world we live in. Light is at frequencies five orders of magnitude higher than 5 GHz. Light is electromagnetic and it does not follow transmission lines. We have a lot to learn as we try to use the next few decades in the frequency spectrum. The answers will lie in physics not circuit theory.

1.3 The Electric Field and Voltage

Before we can talk about moving logic signals, we need to discuss electric and magnetic fields and the definition of voltage. We start our discussion by placing a group of electrons on a small mass. We call this a unit test charge Δq. We then deposit a larger charge Q on a conducting sphere. These charges spread out evenly on the surface of this sphere creating an electric field around the sphere. When the test charge is brought into this field created by Q, the force between the two fields is sensed on the two masses. The repelling force is

$$f = \frac{\Delta q Q}{4\pi r^2 \varepsilon_0} \tag{1.1}$$

where r is the distance to the center of the sphere and ε_0 is a constant called the permittivity of free space. It is a fundamental constant in nature. This equation states that the force between two charges falls off as the square of the distance between the charges. The constant ε_0 is equal to 8.85×10^{-12} farads per meter (F/m). We discuss the farad when we discuss capacitance.

A test charge must be small enough so that the charge distribution on the sphere is not disturbed by its presence. The force per unit charge in the space around the sphere is called an electric or E field. This E field intensity at a radial distance r from the center of the sphere is given by the equation

$$\mathrm{E} = \frac{Q}{4\pi r^2 \varepsilon_0}. \tag{1.2}$$

The E field is called a vector field because it has intensity and direction at every point in space. This force field is proportional to charge and inversely proportional to the square of distance from the center of the sphere. The force direction on the test charge is in the radial direction.

N.B.

Voltage is defined as the work required to move a unit charge between two points in an E or electric field. The points can be in space, between conductors, or between a conductor and a point in space.

Work is simply force times distance, where the direction of the force and the travel direction are the same. If the force is at an angle to the direction of motion, the work equals force times distance times the cosine of the included angle. The work required to move a unit charge from infinity to the surface of a charged sphere is given by Equation 1.3. The simplest path of integration is along a radial line that starts at infinity and goes to the surface of the sphere. Actually, if you do the mathematics correctly, any path will give the same answer. Using Equation 1.1, the voltage V on the spherical surface is

$$\mathrm{V} = \int_{\infty}^{R} f\,\mathrm{d}r = -\frac{Q}{4R\pi\varepsilon_0}. \tag{1.3}$$

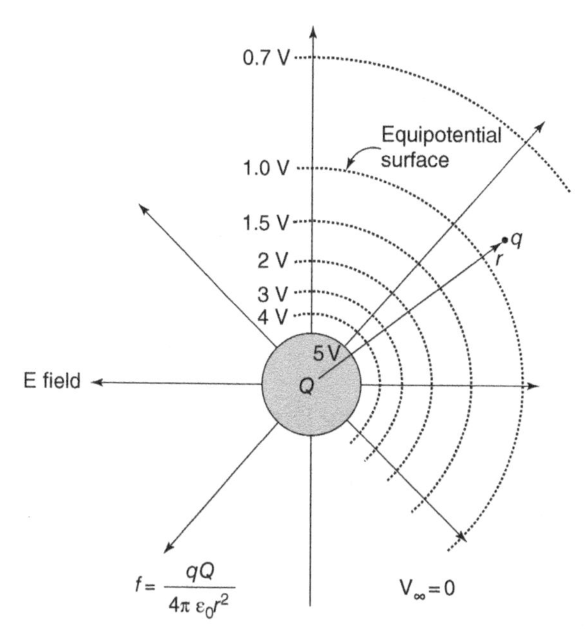

Q is the charge on the sphere
q is small unit test charge
ε_0 is the dielectric constant of free space (permittivity)

Figure 1.2 Equipotential surfaces around a charged sphere.

This equation tells us the amount of work required to move a unit of charge from a large distance to the surface of the charged sphere.

Comparing Equation 1.3 with Equation 1.2, it is easy to see that the intensity of the electric field E at the surface of the sphere is V/R and that E has units of volts-per-meter.

Equation 1.3 can be used to determine the voltage between conductors as well as between points in space. This is shown in Figure 1.2 where the conducting sphere is at a potential of 5 V. If the integration were to stop at 4 V, it would be on an imaginary spherical surface that surrounds the conducting sphere. This sphere is called an equipotential surface. It takes no work to move a test charge on an equipotential surface in space or on the surface of a conductor.

The radial lines represent the force field direction. These lines terminate perpendicular to the conducting surface. If there were a component of the E field parallel to the conducting surface it would imply current flow on the surface. There can be no E field inside of the conducting sphere as this would again imply current flow. At room temperatures, electrons cannot leave the surface of the sphere. Extra electrons space themselves uniformly on the surface to store the least possible amount of energy. This energy is discussed in Section 1.5.

The number of E field lines that we use to describe a field is arbitrary. The intent is to show the shape and relative intensity of the field. One line is assigned to a convenient amount of charge. In a given field pattern, the closer the lines are together, the more intense the field and the greater the force on a unit test charge. In the following diagrams, by convention, lines of force initiate on a positive charge Q and terminate on a negative charge $-Q$. A positive charge is the absence of negative charge. In Figure 1.2 a charge $-Q$ must exist at infinity. In most of the conductor geometries we consider, the E field lines terminate on local conductors. Since these lines represent field shape, their termination at infinity has only symbolic meaning.

1.4 Electric Field Patterns and Charge Distributions

It is convenient to reference voltages to one conductor in a circuit. The reference conductor is assigned the value 0 V. This reference conductor is also referred to as ground or common. There are voltages between conductors which mean there are fields in the intervening spaces. In an analog system, it is possible to have many reference conductors on one board. A reference conductor can be smaller than a nail head or larger than a computer floor.

Reference conductors are often called ground even though they are not earthed. Electrically the ocean could be called a zero of potential but it is not often called ground. A ship's metal hull can be called an electrical ground without creating any confusion.

Figure 1.3 shows the electric field pattern for several conductor configurations that include a large conducting ground plane at the zero of potential. No significant work is required to move charges along this surface or along any connected conductors. Assuming no current flow, the voltage on this ground is constant, independent of the charge distribution on its surface. There is a minor conflict here because there are tangential forces needed to space these charges uniformly apart on the surface. These fields are small and not the subject of this book.

This is not a treatise on free electrons in conductors. The physics of electron motion in conductors involves temperature. There is an average motion of atoms and electrons that is statistical in nature. This average motion in the presence of a field determines the resistivity that varies with each material. For copper, the resistance increases with temperature. For some alloys the temperature coefficient of resistivity is near zero. Predicting the resistance of conductors from basic principles has not been done. Superconductivity occurs at temperatures near absolute zero. The explanation of this phenomena involves quantum mechanics.

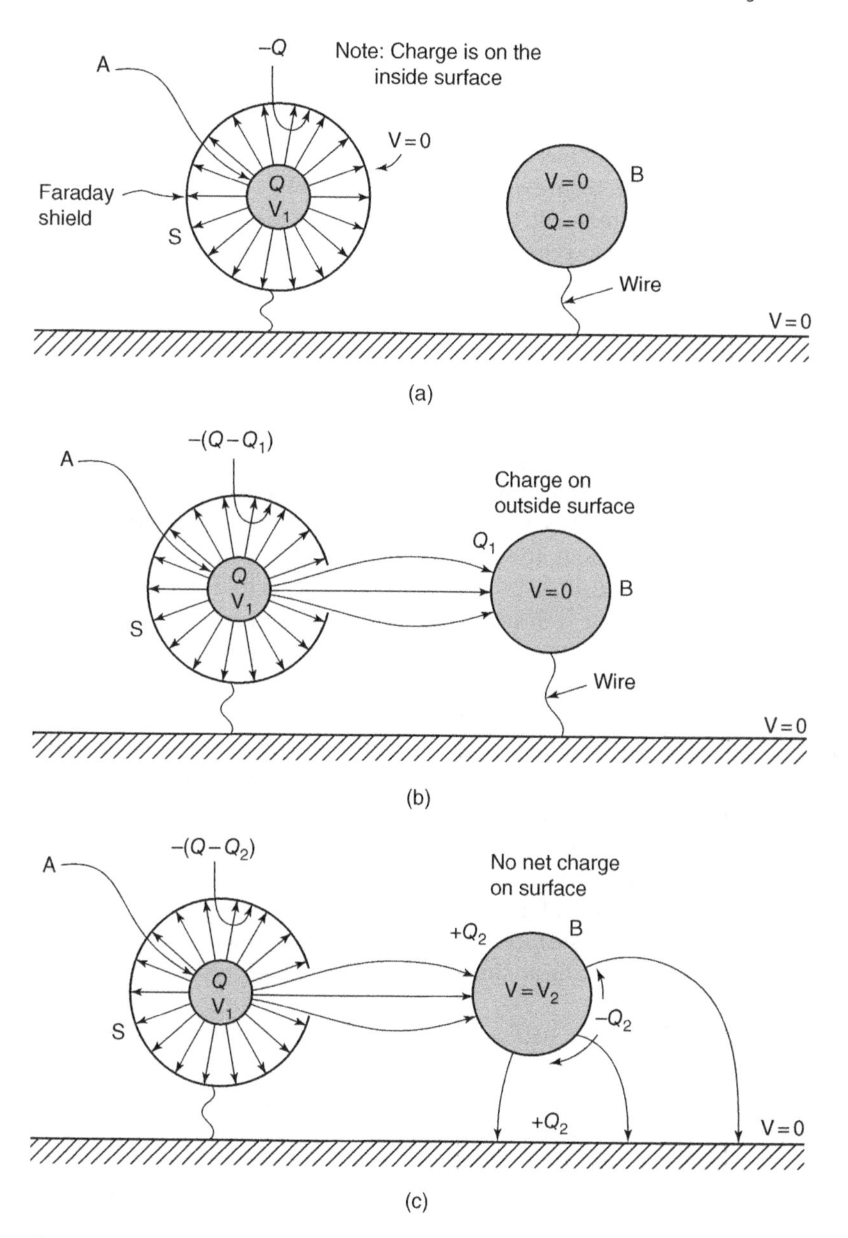

Figure 1.3 (a)–(c) Electric field configurations around a shielded conductor.

INSIGHT

When an attempt is made to measure potential differences between conductors, the probes are conductors that change the very field configuration that is being measured. This problem exists in all measurements.

Jumping ahead for a moment, the voltage difference that is measured along a conductor in a functioning circuit is not an accurate indicator of current flow. A probe cannot distinguish between fields that are inside and outside of a conductor. Only the inner fields cause current flow.

In the field configurations we deal with, the electric field is assumed perpendicular to a conductor's surface even if there is current flow. To illustrate this point, consider the following example. The spacing between a trace and a ground plane is 5 mils or 1.27×10^{-4} m. A trace voltage of 5 V means the E field intensity in the space is about 40,000 V/m. The E field in a trace that supports current flow is usually less than 1 V/m. This ratio tells us that the E field has a very small component that is directed parallel to a trace run. For all practical purposes, the lines representing an E field terminate perpendicular to a conducting surface even if there is current flow.

In Figure 1.3a, a small charged sphere A at potential V_1 is surrounded by a grounded sphere S. No field lines extend outside of this enclosed space. The field lines from A terminate on the inside surface of S. This means a charge $-Q$ has been supplied to this inner surface. The sphere B has no charges on its surface and it is at zero potential. If voltage V_1 changes then charges must be moved from the inside of conductor S to the surface at A. Conductor S is called a Faraday shield. This figure shows that if there is a voltage between conductors, there must be surface charges. There is no exception to this rule. This figure also illustrates a basic shielding principle which is the containment of the electric field.

Figure 1.3b illustrates what happens when a hole is placed in surface S. Some of the field lines from A terminate on conductor B. The result is that a charge Q_1 must be located on surface B. This charge was supplied as current on the wire connecting B to ground. If V_1 changes, then current must flow in the connecting wire to supply a different Q_1. This current is said to be induced. A current flows only when the voltages are changing. If the voltage changes in a sinusoidal manner the current is also a sinusoid.

This figure illustrates several points. If hardware must be shielded, then one hole can violate that shield. A hole is a two-way street. Not only can a field leave the enclosure, an external field can enter through this hole. What is even more important, one lead entering the enclosure can transport interference in either direction. Remember: It only takes one hole to sink a boat.

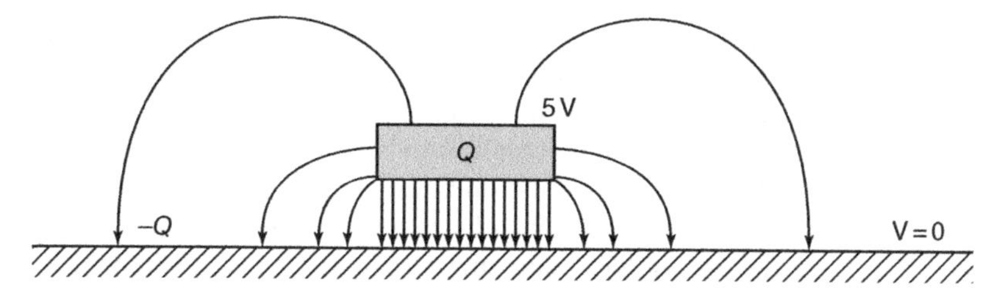

Figure 1.4 The electric field pattern of a circuit trace over a ground plane.

Figure 1.3c illustrates what happens when conductor B is floating. There is no way that new charges can be supplied to this surface, so the sum of the surface charges must be zero. The charges on the surface distribute themselves so that charges $+Q_2$ and $-Q_2$ group on opposite sides of the B surface. If V_1 changes then obviously currents must flow on the surface of B. These are called induced currents. Notice that there is an accumulation of charge on the ground plane under the floating conductor. There are no measureable potential differences along the ground plane. For a system of ideal conductors with changing voltages, surface currents will flow whenever there are terminating field lines in transition. Do not forget that the percentage of charges in motion is one part in a trillion and yet trillions of electrons are in motion. It is fairly obvious that these electrons are truly on the conducting surface. If there were field lines in the conductor there would be current flow.

Figure 1.4 shows the electric field pattern for a circuit trace at voltage V over a conducting plane at 0 V. The field pattern contains a lot of information. The E field intensity is greatest under the trace. Note that the field lines tend to concentrate at sharp edges and that some of the field lines terminate on the top of the trace. The current flow on the surface of a trace will be greatest where the most field lines terminate. This means that the apparent resistance of a conductor depends on how much of the conductor is used for current flow. Areas where the fields do not terminate have no current flow and cannot contribute to conduction. In many applications where high current is involved the conductor geometry must provide mechanical stability. This does not necessarily imply that current will use all of the conductor.

1.5 Field Energy

The work required to move a charge Q up to the surface of a conducting sphere is stored in the electric field around the sphere. There is no other explanation. The temperature is unchanged, there is no mechanical distortion, and there is certainly no chemical change. The same thing happens in the gravitational field. When a pail of water is lifted to a water tank, the gravitational field stores the energy. This concept of space storing field energy is troubling because it defies our senses. We see this movement of gravitational energy in the earth/moon system by observing tidal action. Field energy must also move when a transmission line carries a logic signal. What is actually moving is enigmatic. We discuss energy motion when we discuss transmission lines.

The field in Figure 1.2 extends radially and we know the field intensity at the surface of the sphere. If we consider a plane surface of equal area as in Figure 1.4, we can equate the charge density at the surface with an E-field intensity. Since the E field is a constant over the area A, the increment of work $\mathrm{d}W$ to move a unit charge $\mathrm{d}q$ in this field is $\mathrm{E}h\,\mathrm{d}q$ where E equals $Q/\varepsilon_0 A$,

$$\mathrm{d}W = \mathrm{E}h\mathrm{d}q = \frac{qh}{A\varepsilon_0}\mathrm{d}q. \tag{1.4}$$

The work to move a total charge Q is

$$W = \int_0^Q \frac{qh}{A\varepsilon_0}\mathrm{d}q = \frac{Q^2 h}{2A\varepsilon_0}. \tag{1.5}$$

Since $\mathrm{E} = Q/\varepsilon_0 A$ then $Q = \varepsilon_0 A\mathrm{E}$. Using this value of Q in Equation 1.5 the work W is

$$W = \frac{1}{2}\mathrm{E}^2 V\varepsilon_0 \tag{1.6}$$

where V is the volume of the E field space and W has units of joules. The energy E stored in the volume V is equal to this work or

INSIGHT

$$E = \frac{1}{2}\mathrm{E}^2 V\varepsilon_0. \tag{1.7}$$

Equation 1.7 shows very clearly that space stores electric field energy. Conductors are needed to confine the electric field but they store little energy. If there were E fields in the conductors there would have to be current flow.

There is field energy stored in the spaces defined by any conductor geometry where there are potential differences. It is important to point out that it is impossible to have a static electric field unless there are materials present. When there are insulators (dielectrics) there can be trapped electrons. We will not consider this phenomenon. In Section 1.6, we show that the presence of dielectrics between conductors will change the field patterns and the amount of energy that is stored.

1.6 Dielectrics

Dielectrics are insulators. Glass epoxy is an insulator used in the manufacture of circuit boards. Other dielectric materials are rubber, mylar, nylon, air, polycarbonates, and ceramics. In Figure 1.3, the space between the trace and the conductor is air. If glass epoxy is inserted between the conductors the electric field will reduce in intensity by a factor of about 4. This reduction factor is called the relative dielectric constant ε_R. This factor is simply a number with no units. The voltage between the two conductors will drop to V/ε_R. The E field in Figure 1.4 can now be written as

$$E = \frac{Q}{A\varepsilon_R\varepsilon_0}. \tag{1.8}$$

Insulators and dielectrics play a big role in electronics. Conductors must carry currents between components but they must be supported on insulators, so dimensions are maintained and there is little chance of a voltage breakdown. The characteristics of copper are constant but the insulators that surround copper have continued to evolve. Every insulator is selected for its electrical and mechanical character. Early insulators were paper, Bakelite™, and cloth that have been replaced by glass epoxies, nylon, ceramics, and rubber.

By weight, a circuit board is mostly insulation. As we see in detail, the fields that carry energy use dielectrics and air. The manufacture of circuit boards uses dielectric materials that have proved to be effective. Board designers rely on board manufacturers to select the materials that are best for a design. It is important that board designers stay current with the changing technology. Just another reminder. Energy flows in insulators, not in conductors. The conductors direct where the energy flows. The only energy in the conductors involves moving electrons and this is a tiny fraction of the energy in the space between conductors.

Table 1.1 below lists the relative dielectric constants of some of the common dielectrics and insulators used in electronics.

It is important to note that water has a high dielectric constant. If moisture is absorbed by a dielectric the dielectric constant will change. This means that circuit boards must be coated with a sealant to keep out moisture. There are several sealants available on the market and the choice is related to production quantities and the need to avoid pins, pads, test points, and connections.

Table 1.1 The relative dielectric constant for materials used in electronics.

Material	Relative dielectric constant ε_R
Air	1.0
Barium titanate	100–1,250
BST (barium strontium titanate)	1,000–12,000
Ceramic	5–7
Glass	3.8–14.5
Glass epoxy	3.6–4
Mica	4–9
Mylar—polyethylene terephthalate (PET)	5–7
Neoprene	6–9
Nylon	3.4–22.4
Paper	1.5–3
Teflon	2.1
Water (distilled)	78

1.7 Capacitance

The ability of a conductor geometry to store charge is called capacitance. It is measured as the ratio of charge stored to applied voltage or

$$C = \frac{Q}{V}. \tag{1.9}$$

The unit of capacitance is the farad. A farad stores 1 coulomb of charge for a potential of 1 V. The units in common use are usually below a millionth of a farad. Capacitances as small as 10^{-12} F can be important. This level of capacitance is called a picofarad and is abbreviated as pF. 10^{-9} F is called a nanofarad, abbreviated as nF. 10^{-6} F is called a microfarad, abbreviated as μF.

For the sphere using Equation 1.3, the ratio of charge to voltage is

$$C = 4\pi r \varepsilon_R \varepsilon_0. \tag{1.10}$$

Using Equation 1.10, the capacitance of the earth is 711 μF.

For parallel conducting planes the capacitance is

$$C = \frac{\varepsilon_R \varepsilon_0 A}{h} \tag{1.11}$$

where A is the surface area and h is the spacing between conductors. This equation can be used to calculate the capacitance of most commercial capacitors. It can also be used for determining the capacitance between conducting planes or the capacitance of a trace over a conducting plane. Assume that a 10-mil wide, 10-cm long trace is 5 mils above a ground plane. If the relative dielectric constant is 4 the capacitance is 7.1 pF.

1.8 Capacitors

A capacitor is a conductor geometry that stores electric field energy. There are many types of commercial capacitors that fill a wide range of needs. They can act as filters in power supplies. They can function to block dc in analog circuits. They can act as elements in active and passive filters, function generators, and oscillators. They come in many shapes and sizes and can handle a wide range of voltages. They can weigh a ton or be as small as a pin head. Depending on application they are constructed using many different dielectrics. Most of the capacitors we deal with in this book are called decoupling capacitors. They supply energy to transmission lines and semiconductor components. These capacitors are surface mounted which means they have no leads. The character of these capacitors will be examined in more detail later in this book.

The energy stored in a capacitor is given by Equation 1.12 where V is the voltage, A is the area, and h is the conductor spacing. The energy is thus

$$E = \frac{AV^2}{2h}\varepsilon_0\varepsilon_R. \tag{1.12}$$

Substituting Equation 1.11 into Equation 1.12, the energy equals

$$E = \frac{1}{2}CV^2. \tag{1.13}$$

The capacitors used on a logic circuit board typically use ceramic dielectrics. The dielectric is often BST (Barium Strontium Titanate). This dielectric material can be formulated to have a dielectric constant as high as 12,000. This high dielectric constant allows the capacitors to be both small and inexpensive. Most circuit boards use dozens of these capacitors. The dielectric constant varies with temperature and voltage. Since the exact capacitance is not critical, this type of capacitor is ideal for supplying energy to transmission lines on a circuit board. The function of these capacitors is discussed in Chapters 2 and 3.

1.9 The D or Displacement Field

It is necessary to have two measures of the electric field. The E or electric field is known when voltages are known. A second measure of the electric field is the D field. Lines of D field start and stop on charges. The D field does not change

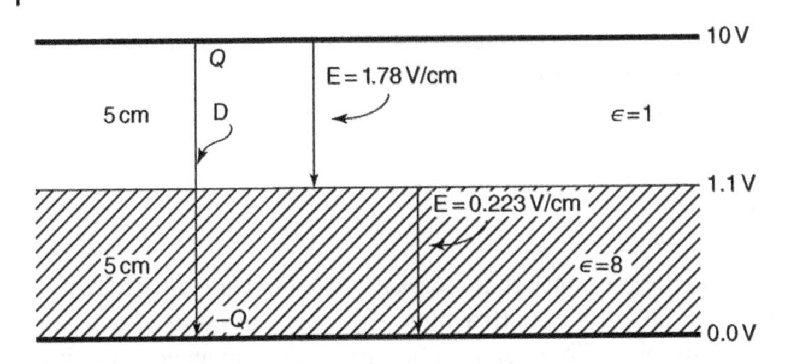

Figure 1.5 The electric field pattern in the presence of a dielectric.

value across a charge-free dielectric boundary. The D field allows us to extend the concept of current flow to space. The D field is known when the charge distribution is known. In our discussions, there will be no free charges in space or at dielectric boundaries. There will only be a very small percentage of available charges moving on the surface of conductors.

The D field is known as the displacement field and it is related to the E field by the permittivity of free space and the relative dielectric constant

$$D = \varepsilon_0 \varepsilon_R E. \tag{1.14}$$

In Figure 1.5, the D field is continuous between the two conductors. It does not change at the dielectric boundary. The voltage across the two dielectrics must sum to 10 V. The E field in the dielectric is 1/8 the E field in air. The voltage across the dielectric is (E/8) × 5 cm and across the air space is (E/1) × 5 cm. The sum is 10 V. The E field in free space is 1.78 V/cm and 0.233 V/cm in the dielectric. Note that most of the field energy is stored in the air space.[2]

1.10 Mutual and Self Capacitance

A mutual capacitance is often referred to as a leakage or a parasitic capacitance. On a circuit board where traces run parallel to a conducting plane, some of the field lines will terminate on nearby traces rather than on the plane. This situation is shown on Figure 1.6.

The self-capacitance of a conductor is the ratio of charge to voltage on a conductor when all other conductors are at zero potential. The ratio of charge on conductor 2 to voltage on conductor 1 with all other conductors grounded is called a mutual capacitance C_{12}. If the charge on conductor 1 is positive, the

2 High voltage transformers are often immersed in a liquid dielectric to limit the E field near conductors and thus control arcing. The liquid also helps conduct heat out of the transformer core.

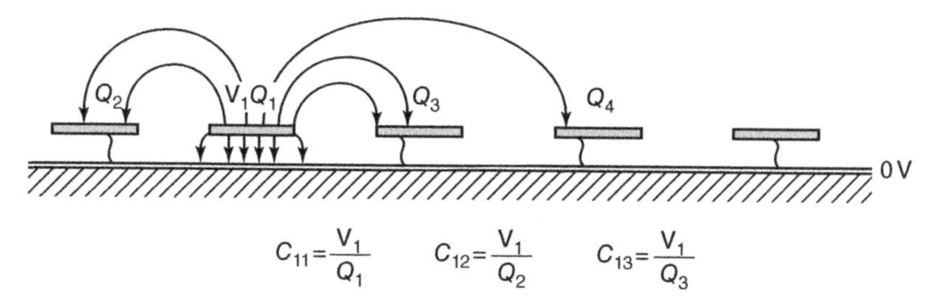

Figure 1.6 The mutual capacitances between traces over a ground plane.

charge on conductor 2 is negative. Therefore, all mutual capacitances are negative. It can be shown that C_{12} is equal to C_{21}.

On most circuit boards with short trace runs, mutual capacitances can be ignored. For long trace runs, the cross-coupling that results can impact signal integrity. This subject is discussed in Section 4.5.

1.11 Current Flow in a Capacitance

In circuit theory, the signals of interest are sine waves. In logic, the ideal signals of interest are step functions of current or voltage. If a steady current I flows into a capacitance C the voltage will increase linearly. The charge that flows is the current in amperes times time or

$$Q = I \times t. \tag{1.15}$$

From Equation 1.9, $Q = CV$. This means that

$$\mathrm{V} = \frac{I \times t}{C}. \tag{1.16}$$

The voltage across a capacitor, when supplied a steady current, is shown in Figure 1.7.

Mathematically, the current I that flows into a capacitor when the voltage changes with time is

$$I = C\frac{\mathrm{dV}}{\mathrm{d}t}. \tag{1.17}$$

A steady current flowing into a capacitor causes the voltage to rise linearly. The current flow accumulates a charge on the capacitor plates at a steady rate. In circuit theory, a surface charge is not considered and the current flow is assumed continuous through the capacitor. In field theory, a changing displacement or D field is equated to current flow. The changing D field means there are changing charges present on the plates of the capacitor. This concept will be important in the study of waves on transmission lines.

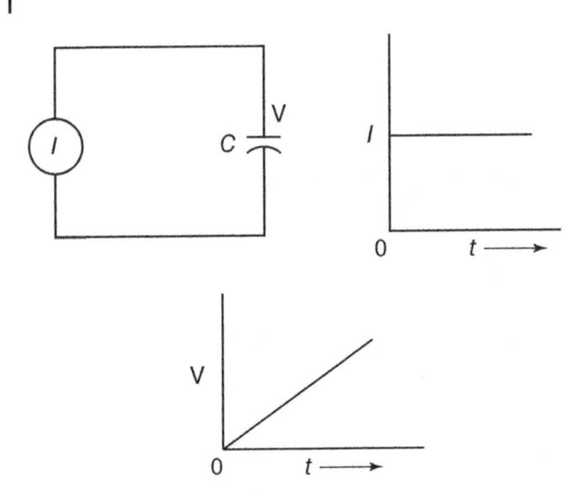

Figure 1.7 The voltage on a capacitor when supplied a steady current.

N.B.

The essence of the electric field is simple. A voltage difference implies an E field. A changing D field is current flow in space. The relation between the two representations of the electric field is the dielectric constant.

In circuit theory, the current that flows in a capacitance is related to reactance. This avoids relating current flow to a changing D field.

1.12 The Magnetic Field

Magnetic fields and electric fields play an equal role in moving energy. Before we can discuss moving energy, we need to spend some time understanding magnetic fields. When we observe logic signals we look at voltages. The presence of voltage immediately implies an electric field. What may not be appreciated is that moving a voltage between two points requires moving energy and this requires both a magnetic and an electric field. When energy is not moving, the magnetic field is usually zero.

An understanding of magnetic fields is required in the design of transformers, speakers, inductors, motors, generators, MRI hardware, and particle accelerators. Magnetic fields are also required in moving logic signals on traces. Magnetic fields are a part of all electrical activity including electromagnetic radiation. Electromagnetic field energy in motion is always divided equally between electric and magnetic fields. Half the field energy in sunlight is magnetic.

We have all played with permanent magnets and sensed forces at a distance. We can duplicate these same forces when current flows in a coil of wire. A very simple experiment can show the shape of a magnetic field. A wire is passed through a piece of paper holding iron filings. When a steady current flows in the wire, the filings will line up forming circles around the wire. These closed circles show the shape of the magnetic field. The filings line up to minimize the amount of energy stored in the magnetic field.

The atomic structure of some elements allows their internal fields to align. In these materials, we can observe magnetic behavior. Common magnetic materials include iron, cobalt, and nickel. The rare earths samarium, dysprosium, and neodymium are also magnetic. These elements, when properly blended and annealed, provide permanent magnets with very strong magnetic fields. Uses include speakers in our phones and computers. They make electric motors in battery-powered automobiles practical.

The iron in the earth's core is magnetized and forms a huge magnetic field that surrounds the earth. A compass needle will align with this magnetic field to indicate direction on the surface of the earth. The earth's magnetic field concentrates near the earth's magnetic poles. The Aurora Borealis in the northern hemisphere is caused when electrons that arrive from the sun spiral in the earth's magnetic field and concentrate near the earth's north magnetic pole. These electrons ionize air molecules in the upper atmosphere causing radiation in the visible part of the spectrum. In the southern hemisphere the effect is called Aurora Australis.

When current flows in a conductor, there is a magnetic field. When a second conductor carrying current is brought close, there is a force between the two conductors. If the current flows in the same direction, the conductors will attract each other. The attraction and repulsion of magnetic fields is at the heart of motors and generators. This same magnetic field is involved in moving a logic voltage on a circuit board and that is the focus of our attention.

The force between two conductors carrying current is perpendicular to the direction of both the magnetic field lines and the current flow. A magnetic field has direction and intensity at every point in space and is therefore a vector field. Unlike the electric field, magnetic fields are represented by closed curves. The simplest patterns are the field circles around a single straight conductor carrying current as shown in Figure 1.8.

Ampere's law states that the line integral of the H field around a conductor is equal to the current I in the conductor. In Figure 1.8, the H field is constant at a distance r from a long conductor which means that $2\pi r\mathrm{H} = I$. The magnetic field intensity H at a distance r from a long conductor carrying a current I is

$$\mathrm{H} = \frac{I}{2\pi r} \tag{1.18}$$

where H has units of amperes per meter. Just like the electric field, the magnetic field shape is characterized by a set of curves. Just as in the E field, when the lines are close together, the field intensity is greatest.

An insulated conductor wound in a coil is called a solenoid. This arrangement is shown in Figure 1.9. The H field intensity in the center of the coil is

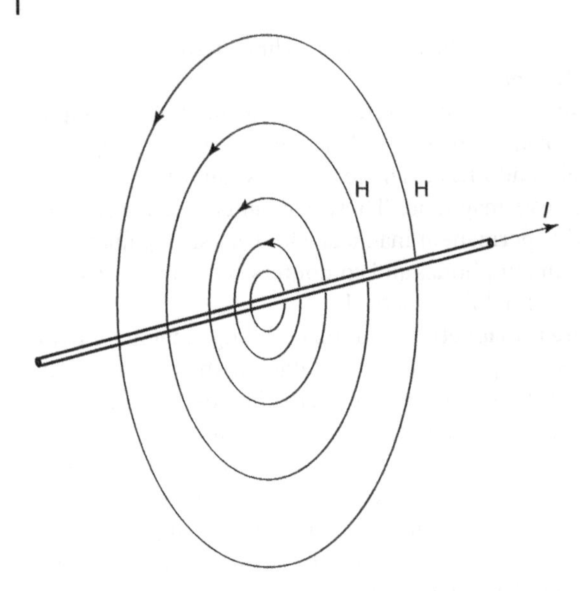

Figure 1.8 The magnetic field H around a current carrying conductor.

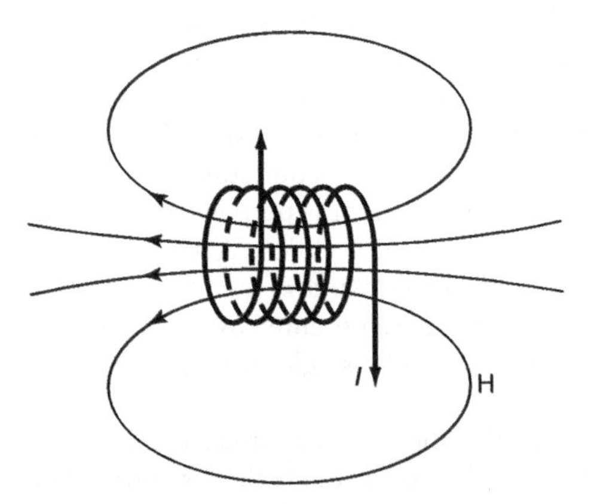

Figure 1.9 The H field in and around a solenoid.

nearly constant and it is proportional to the current and the number of turns. The field intensity in the coil is given approximately by

$$H = \frac{In}{\ell} \tag{1.19}$$

where I is the current in amperes, n is the number of turns, and ℓ is the length of the coil.

There is no element of magnetic material that corresponds to the electron in the electric field. This means there is no magnetic pole that can be moved in the field to add energy. To understand the work required to establish a magnetic field, we need to discuss the concept of inductance and the induction or B field.

1.13 The B Field of Induction

In magnetics, we use an H field that is proportional to current flow and a B field that relates to induced voltage. This parallels the case in electrostatics where we described the electric field as a D field that originated on charges and as an E field that related to voltage. In electrostatics, the E field is the force field and in magnetics the B field is the induction or force field. In the electric field the presence of a dielectric reduces the E field and the energy density. In magnetics, the H field intensity is reduced in the presence of magnetic materials. Lines of force in the B field are called magnetic flux lines. The B field has units of teslas. B field intensity does not change at a boundary where the permeability changes. The H field is often called a magneto-motive force. In a material with high permeability, it takes very little H field to establish a B field. In air, it takes a high magneto-motive force to establish the B field.

When a changing B field flux couples to an open coil of wire, a voltage will appear at the ends of the coil. This is shown in Figure 1.10. A voltage can result if the coupled flux changes when an open coil is rotated or moved in the field. If the number of flux lines remains unchanged, the voltage is zero.

When there is a current there is an H field in space. At every point in space the H field has an associated B field given by

$$\mathrm{B} = \mu_0 \mu_\mathrm{R} \mathrm{H} \tag{1.20}$$

where μ_0 is the permeability of free space and μ_R is the relative permeability. The value of μ_0 is $4\pi \times 10^{-7}$ Tm/A (tesla meters per ampere). This a fundamental constant in nature. The voltage induced in a coil is

$$\mathrm{V} = nA\frac{\mathrm{dB}}{\mathrm{d}t} \tag{1.21}$$

where A is the area of the loop and n is the number of turns and $\mathrm{dB}/\mathrm{d}t$ is the rate the B flux changes with time. This equation is known as Faraday's law. This

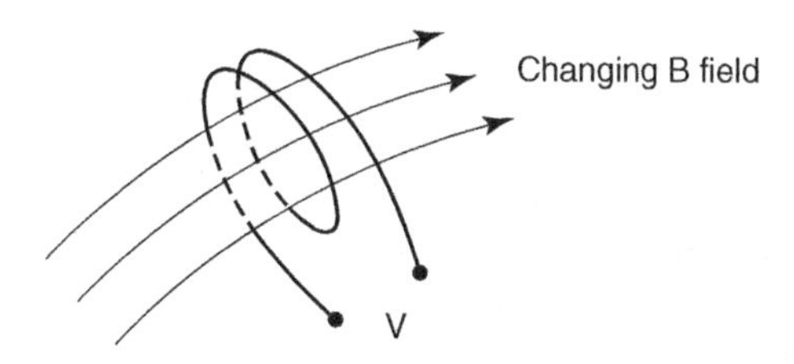

Figure 1.10 A voltage induced into a moving coil.

equation states that if the coupled induction field B changes intensity at a fixed rate, a steady voltage will appear at the coil ends. It also says that if a steady voltage is applied to the coil ends, the B field must increase linearly with time. This equation is true even if the coil is in air. Generally, at low frequencies, magnetic materials are used in the magnetic path to limit the magnetizing current needed to establish the B field.

In this book, we discuss the magnetic fields that surround traces carrying current. There will be no magnetic materials present, but there may be cross-coupling between traces. Currents in traces will create an H field. Cross-coupling between traces will involve the induction or B field. It will also turn out that the energy stored in a magnetic field involves both the B and H measure of the field.

N.B.

All the discussion above can be stated as follows: A changing B field generates voltage and a current generates an H field.

1.14 Inductance

The definition of inductance is the ratio of magnetic flux to current flow or

$$L = \frac{\varphi}{I}. \tag{1.22}$$

As an example, the flux φ in a solenoid is equal approximately to the B field times the cross-sectional area of the coil.

In general, exact equations of inductance using this definition are very difficult to generate. The inductance of a conductor geometry can be measured by noting the voltage that results when current changes value or

$$V = L\frac{dI}{dt}. \tag{1.23}$$

From this equation, it is obvious that every conductor geometry that can carry current generates magnetic flux and therefore has inductance. This includes the current flow in a capacitor. In circuit theory, magnetic fields are generally ignored. In transmission line theory, the magnetic field associated with a changing electric field is fundamental to moving energy.

Inductors are components that are designed to store magnetic field energy. At frequencies below a MHz, inductors usually require magnetic materials to direct the magnetic field to a volume of space called a gap. In later chapters, the inductance we consider involves traces running over plane conductors and involves no magnetic materials. The energy moved and stored in this inductance is key to logic operations.

There is magnetic field energy stored between a trace and a plane conductor when current flows in this loop. There is an electric field in this same space when a voltage is placed between the trace and a conducting plane. A trace over a conducting plane is thus a combination of an inductor and a capacitor. It is conventional to call this conductor geometry a transmission line.

The unit of inductance is the henry. Components that are intended to store magnetic field energy are called inductors. The inductance of typical components can range from microhenries (μH) to henries (H). When we study transmission lines, parallel conductors will have an inductance per unit length measured in nanohenries (nH).

Energy can be entered into an inductor by placing a voltage across its terminals. The power supplied at any moment is the product of voltage and current. The energy stored in the inductor is the integral of power over a time t. The energy is

$$E = \int_0^t VI\,\mathrm{d}t. \tag{1.24}$$

Substituting Equation 1.23 for V

$$E = \int_0^I LI\,\mathrm{d}I = \frac{1}{2}LI^2. \tag{1.25}$$

In Figure 1.7, when a constant voltage is applied to an inductor, the current ramps up in a linear manner. This is shown in Figure 1.11.

An isolated conductor has inductance. The magnetic flux generated by current flow depends on conductor length and only slightly on conductor diameter. Large

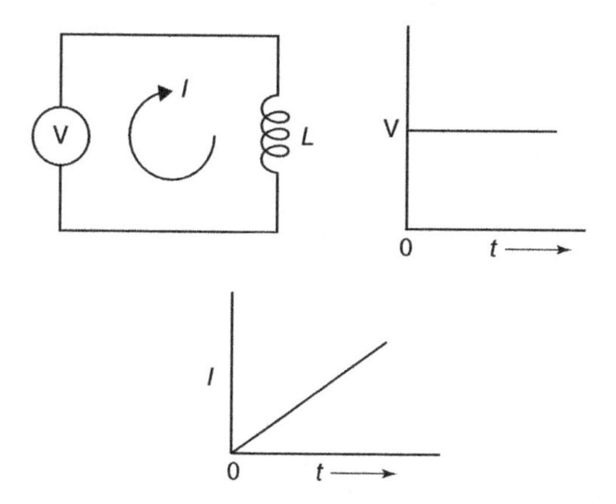

Figure 1.11 An inductor driven from a constant voltage source.

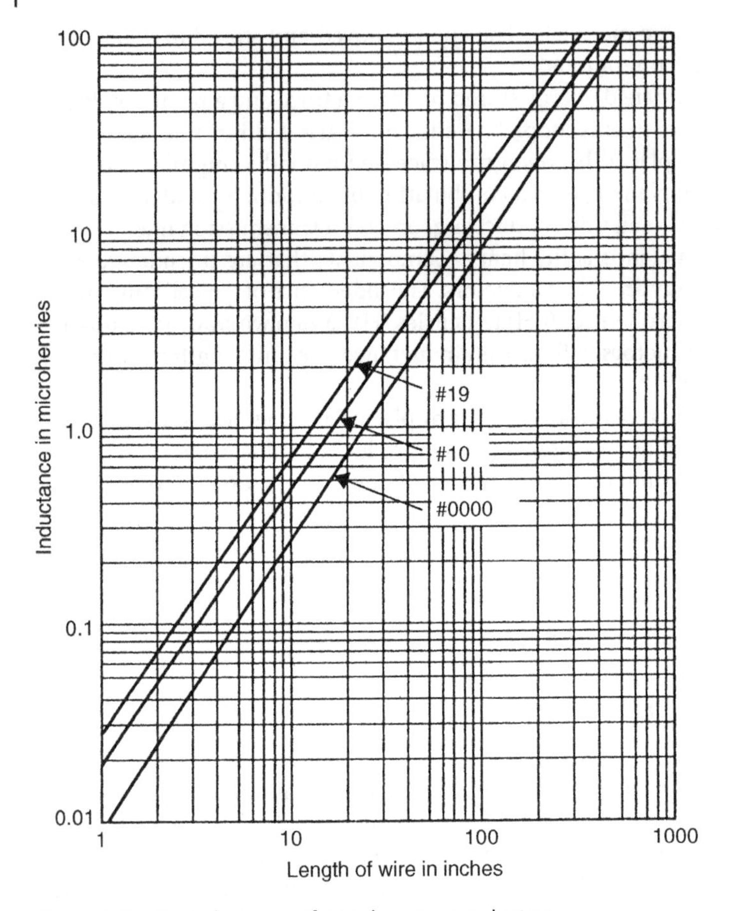

Figure 1.12 The inductance of round copper conductors.

diameters are required if high currents are involved. In a building, the I beams are in effect inductors that modify the fields that enter or are generated in the area. These fields are also modified by equipment racks, ground planes, conduit, and all metal surfaces. Attempting to reduce potential differences by "shorting" points together will usually fail. If an I beam did not affect the field, how can a no. 10 conductor be effective? The inductance of a length of round conductor is shown in Figure 1.12.

1.15 Inductors

An inductor is a component that stores magnetic field energy. The inductors we consider later are made using trace geometries on a circuit board. These inductances are effective at high frequencies and usually do not involve magnetic materials. Inductances in the range 50 nH can be used to match transmission lines to a transmitting antenna at hundreds of MHz.

At lower frequencies, inductances can be supplied as components. These components can be turns of copper in air or turns wound around a magnetic core. When magnetic materials are involved, they shape the field but the energy is still stored mainly in air. In larger inductors, the magnetic flux crosses through air in a controlled gap. This is where the energy is stored.

A magnetic material in common use in circuit components is called ferrite. A ferrite core consists of small bits of magnetic material embedded in a ceramic filler that when fired forms a hard insulator. As an insulator, eddy current losses are limited to small islands of conductive material. Ferrite is the only magnetic material that has a useful relative permeability above 1 MHz. A typical gap structure is shown in Figure 1.13.

Cup cores can be used to build an inductor. This core construction is shown in Figure 1.13. The core surfaces that touch are ground flat and polished. A controlled gap can be provided in the center of the core by grinding the center contacting surface. For transformer action, no gap is provided. This type of construction is shown in Figure 1.14.

A single turn threading a small ferrite bead forms an inductor. The reactance at 1 GHz is theoretically under an ohm and this assumes there is permeability at this frequency. Using a bead as a filter element to reduce noise is usually ineffective and is not recommended. The best approach in design is to avoid reflections, not add them. Often the role of the bead is to

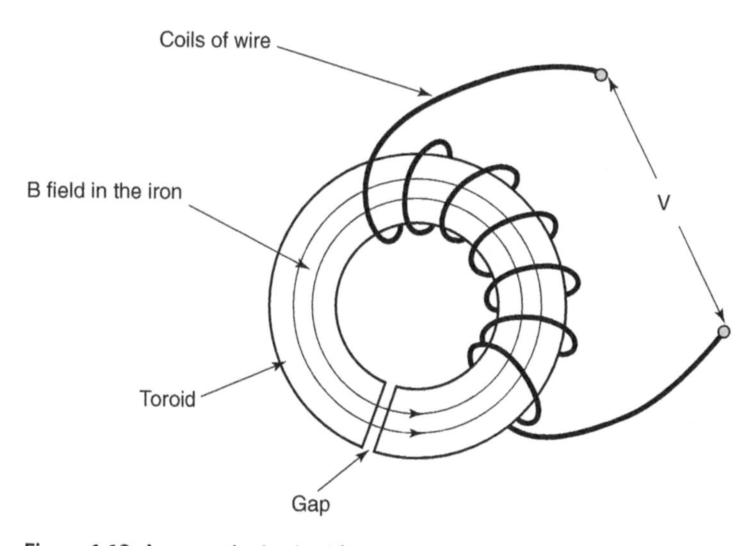

Figure 1.13 A magnetic circuit with an air gap.

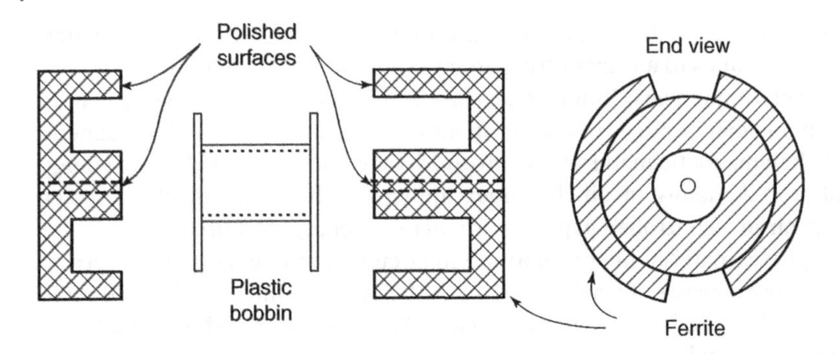

Figure 1.14 Ferrite cup core construction.

space conductors which reduce cross-coupling. This can be done without adding a component.

1.16 The Inductance of a Solenoid in Air

A solenoid is a multiturn coil much like the symbol for inductor. The solenoids we are interested in consist of a few turns made from traces and vias. Magnetic materials are not considered as they are ineffective at the frequencies of interest.

From Equations 1.18 and 1.19, the B field in a solenoid is

$$\mathrm{B} = \frac{\mu_0 I n}{\ell}. \tag{1.26}$$

From Equations 1.21 and 1.26

$$\mathrm{V} = nA\frac{\mathrm{dB}}{\mathrm{d}t} = \frac{n^2 A\mu_0}{\ell}\frac{\mathrm{d}I}{\mathrm{d}t}. \tag{1.27}$$

Referencing Equation 1.23, the self inductance of a solenoid is approximately

$$L = \frac{n^2 A\mu_0}{\ell} \tag{1.28}$$

where A is the area of the coil, ℓ is the length of that coil, and n is the number of turns. The dimensions are in meters, $\mu_0 = 4\pi \times 10^{-7}$ tesla meters per ampere and the inductance is in henries.

An inductance of 10 nH has a theoretical reactance of 6.28 ohms at 100 MHz. It is not possible to build an inductor with zero parasitic capacitance. The natural frequency that results places an upper limit to what might be considered a simple inductor. It would be safe to say that at 1 GHz there are no simple components.

An imbedded inductor in the form of a solenoid can be formed on a circuit board using vias and traces. If the coil pitch is 15 mils where a single turn is 60 mils square, 10 turns will have an inductance of 76 nH. An increased pitch and more turns will help to increase the natural frequency.

1.17 Magnetic Field Energy Stored in Space

To calculate the magnetic field energy in a volume of space, consider a trace over a conducting plane as shown in Figure 1.15.

If the spacing h is small, most of the field energy will be stored in the volume under the trace. Ampere's law requires that line integral of H around the current in the trace is equal to the current I. In this figure, most of the current flows out on the underside of the trace and returns under the dielectric on the conducting plane. Assuming that the only part of the path that contributes to the integral is along the width w under the trace,

$$\int \mathrm{H} dw = I. \tag{1.29}$$

If we assume H is uniform under the entire trace then

$$\mathrm{H} = \frac{I}{w}. \tag{1.30}$$

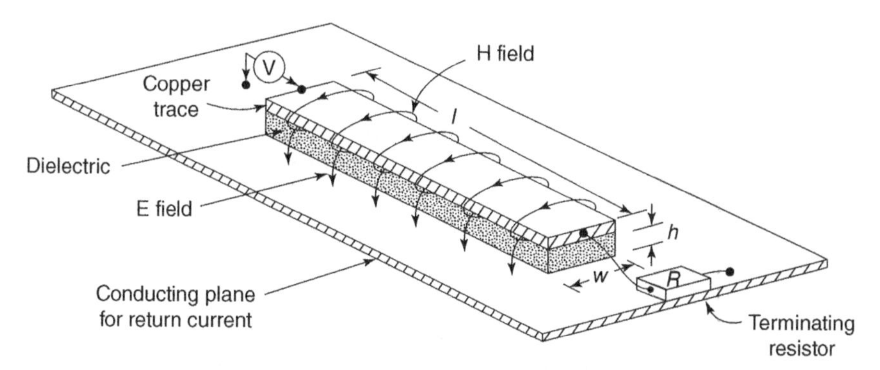

Note: In logic, the forward current flows mainly on the underside of the trace and returns under the dielectric on the conducting plane.

Figure 1.15 A trace over a conducting plane showing fields.

The magnetic B flux in an area $h\ell$ is

$$\varphi = \mathrm{B}\ell h. \tag{1.31}$$

Substituting Equations 1.22 and 1.28 into Equation 1.25, the energy in the magnetic field is

$$E = \frac{1}{2}\varphi I. \tag{1.32}$$

Since $I = \mathrm{H}w$, $\varphi = \mathrm{B}\ell h$, and the product ℓhw is volume V, the energy in the field is

$$E = \frac{1}{2}\mathrm{BH}V. \tag{1.33}$$

In magnetic material, the H field is very small and there is little energy storage. In a volume of space

INSIGHT

$$E = \frac{1}{2}\cdot\frac{\mathrm{B}^2 V}{\mu_0} \tag{1.34}$$

Equation 1.32 shows very clearly that space stores magnetic field energy. Just as in the electric field, it takes a conductor geometry to contain this field. In the case of the capacitor, energy can be stored for long periods of time. For inductors, sustained energy requires a sustained current. This is possible when a superconductor is involved.

1.18 Mutual Inductance

The current flowing in one conductor geometry can generate magnetic flux in a second conductor geometry. The ratio of this flux to the initial current is called a mutual inductance.

$$L_{21} = \frac{\varphi_2}{I_1}. \tag{1.35}$$

This changing flux can cross-couple an interfering voltage into a nearby circuit. This coupling is most likely to happen when transmission lines run in parallel over a distance. Unlike mutual capacitance, mutual inductance can be of either polarity.

This coupling is discussed in Section 4.5.

1.19 Transformer Action

It is often very useful to generate multiple dc voltages to operate logic. Dc–dc converters can provide several voltage levels that can be referenced to points on a board. Converters are available as components.

One method of adding needed operating dc voltages uses transformer action. A square-wave voltage is connected to a coil wound on a magnetic core without a gap. The voltage creates a steadily increasing B field. The B field intensity in iron depends only on voltage, the number of turns, the area of the core, and the time the voltage is connected. If a secondary coil is wound over the first coil, the magnetic flux will couple to this second coil. If the ratio of turns is 4 : 1, the voltage on the secondary coil is reduced by a factor of 4.

The magnetic flux in a core increases as long as a voltage is placed across a primary coil. When the flux level is near the permitted maximum, the voltage is reversed in polarity and the flux intensity starts to reduce. In time, the flux level goes through zero and reaches its negative limit. At this time the voltage polarity is again reversed. In a core with high relative permeability, the current that is needed to establish the H field is very small. This assumes there is no gap in the core. If a load is placed on the secondary coil the current that flows will obey Ohm's law. The primary current will increase to supply the required energy into the load. The secondary voltages must be rectified, filtered, and regulated to serve as a dc power supply.

1.20 Poynting's Vector

The idea that energy is moved in coupled electric and magnetic fields is fundamental in nature. The practical problem we face is that there are no direct tools for measuring energy flow or energy storage in space. The parameters we can measure are voltage, current, capacitance, and inductance. Poynting's vector is a statement of power flow at a point in space using field parameters—not circuit parameters. This vector further presses the point that energy flows in space and that the conductors direct the path of energy flow. This vector is also valid in free space where there are no conductors. Figure 1.16 shows Poynting's vector applied to a simple transmission line. The vector is mathematically the cross-product of the **E** and **H** vectors or

$$\mathbf{P} = \mathbf{E} \times \mathbf{H}. \tag{1.36}$$

The total power crossing an area is the integral of the Poynting's vector component that is perpendicular to that area. It will be shown later that Poynting's vector does not exist behind every wave front. Some waves convert field energy while others carry field energy.

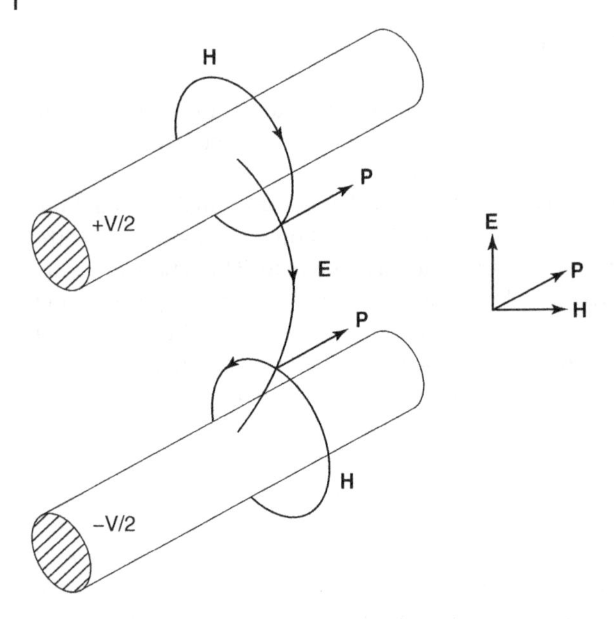

Figure 1.16 Poynting's vector for parallel conductors carrying power.

Poynting's vector has no frequency limits and applies to dc fields. The term dc implies an unchanging field in a region of space over a period of time. Utility power is carried in slowly changing fields in the space between conductors. A flashlight carries power from the battery to a lamp in space. A transmission line carries energy from a decoupling capacitor to a logic gate in the space between a trace and a ground plane. Space carries energy. Conductors direct where the energy travels.

1.21 Resistors and Resistance

Circuit theory is about resistors, capacitors, and inductors and how they function in networks at different frequencies. On a logic circuit board, resistors play a secondary role by terminating transmission lines. Resistance in a parasitic sense plays a role in dissipating the energy needed to move signals on these lines. This energy is dissipated in logic switches as well as in the traces that carry current. Initial currents stay on the surface of conductors. This effect means that most of the copper is not used and this raises the effective resistance. This is called skin effect and will be discussed later in the book.

The resistors that are used in analog work cover the range from 10 ohms to perhaps 100 megohms. All resistors have shunt parasitic capacitances of a few picofarads and the leads have an inductance of a few nanohenries. Circuit designers recognize these parasitic effects and design accordingly. In logic design, a typical impedance level is 50 ohms. This impedance is nominal whether the clock rate is 1 MHz or 1 GHz. This value is bounded above by the impedance of free space which is 377 ohms and the fact that it is impractical to work with 5-ohm circuits. It is interesting that the geometric mean between 5 and 500 ohms is exactly 50 ohms. You may have noticed that over the years the characteristic impedance of transmission lines has remained constant even though clock rates have risen.

All resistors have a shunt parasitic capacitance. The shunt capacitance can be lowered by reducing the diameter and increasing the length of the resistor. Even the lead diameter must be considered. In this text, the resistors that are mentioned are ideal. To deal with a network that represents a practical resistor in a real environment is not useful. What is important is that compromise is necessary and being too idealistic will make it impossible to proceed.

The losses in conductors play a role in carrying a logic signal to its destination. If this resistance were zero, then the losses would be limited to the logic switches. As you will see, this energy cannot be saved or put back into the power supply. There is a good argument for limiting line losses as this maintains signal integrity. At the same time this puts a greater burden on the logic switches to dissipate this energy. A perfect switch cannot dissipate energy. Energy that is not dissipated or radiated must oscillate between storage in inductance and capacitance. All conductor geometries have capacitance and inductance.

The general engineering problem involves the resistance of conductors. If a trace is solder plated the surface resistance is raised and the losses increase. If a trace is silver plated the surface resistance drops and the losses drop. If a trace is treated with a conformal coating the current will use the copper. It is important to realize that there is no way to avoid dissipating energy. The resistivity ρ of various conductors is shown in Table 1.2. This table assumes the entire conductor is carrying current.

Table 1.2 The resistivity of common conductors.

Conductor	Resistivity (μΩ cm)
Aluminum	2.65
Copper (annealed)	1.772
Gold (pure)	2.44
Iron	10
Lead	22
Nichrome	100
Platinum	10
Silver	1.59
Lead-free solder	12
Tin	11.5

The resistance of a conductor assuming an even flow of current is

$$R = \frac{\rho \ell}{A} \tag{1.37}$$

where ℓ is the conductor length and A is the cross-sectional area.

If the strip is a square where the width is equal to the length, the resistance is equal to

$$R = \frac{\rho}{t}. \tag{1.38}$$

Thus, a square of material has the same resistance regardless of its size. The resistance in units of ohms-per-square depends only on thickness. This assumes that the current flow is uniform across the square and uses all the conducting material. The resistance does increase with frequency as skin effect limits the current penetration (see Section 2.17).

The resistance of copper and iron squares for moderate frequencies is shown in Table 1.3. This table is for sine waves, not step functions. Skin effect dominates the resistance for logic where most of the current flows on one surface.

It is interesting to apply this table to a lightning pulse where the current flows uniformly across a square of copper. Assume the square has a resistance of 300 micro-ohms per square. A current of 100,000 A would cause a voltage drop of 30 V. For a 0000 conductor the inductance of a 10-foot section would be about 10 μH. The voltage drop for a lightning pulse would be about 4 million volts (see Figure 1.8). Connecting surfaces together using large diameter conductors is ineffective. To limit voltage drops, surfaces should be connected together so that current flow is not concentrated.

Table 1.3 Ohms-per-square for copper and iron.

	Copper			Steel		
Frequency	t=0.1 mm	t=1 mm	t=10 mm	t=0.1 mm	t=1 mm	t=10 mm
10 Hz	172 μΩ	17.2 μΩ	17.2 μΩ	1.01 mΩ	101 μΩ	40.1 μΩ
100 Hz	172 μΩ	17.2 μΩ	3.35 μΩ	1.01 mΩ	128 μΩ	126 μΩ
1 kHz	172 μΩ	17.5 μΩ	11.6 μΩ	1.01 mΩ	403 μΩ	400 μΩ
10 kHz	172 μΩ	33.5 μΩ	36.9 μΩ	1.28 mΩ	1.26 mΩ	1.26 mΩ
100 kHz	175 μΩ	116 μΩ	116 μΩ	4.03 mΩ	4.00 mΩ	4.00 mΩ
1 MHz	335 μΩ	369 μΩ	369 μΩ	12.6 mΩ	12.6 mΩ	12.6 mΩ
10 MHz	1.16 mΩ	1.16 mΩ	1.16 mΩ	40.0 mΩ	40.0 mΩ	40.0 mΩ

Problem Set

1. Assume a grain of sand is 0.1 mils in diameter. How many grains are there in a cubic foot?
2. Assume the current on a copper trace penetrates uniformly to a depth of 0.2 mils. What is the resistance per meter of a copper trace 5 mils wide?
3. A 50-ohm resistor has a parasitic capacitance of 2 pF. What is the frequency where the reactance falls to 45 ohms? Use admittances.
4. The capacitance of a trace is 2 pF/cm. The relative dielectric constant is 4. What is the inductance per cm if the characteristic impedance is 50 ohms?
5. What is the current flow in a capacitance of 100 pF if the voltage across the capacitance changes 5 V in 2 ns?
6. The charge on an electron is 1.602×10^{-19} coulombs. How many electrons pass a point per second for a current of 100 mA?
7. Now many free electrons are in a gram of copper (about 1 cm of trace)?
8. Assuming only the top 0.01% of the copper contributes to current flow, what is the electron velocity?
9. What is the voltage across an inductor of 1 nH if the current rises 100 mA in 2 ns?

Glossary

Ampere. A flow of charge at the rate of one coulomb per second. It is also the flow of displacement current represented by a changing D field.

Ampere's law. The H field around a long current carrying conductor is equal to $I/2\pi r$. The field lines are circles. The law stated in mathematical terms is: The line integral of H around any closed path is equal to the current passing through that loop (Section 1.12).

Avogadro's number. The number of atoms in a gram-mole of any element. The number is 6.02×10^{23}. For copper the number of grams in a mole is 63.54 (Section 1.2).

B field. The B field is the field of magnetic induction. The intensity of the field at a point is measured in teslas. A B field is represented by lines that form closed curves that do not change intensity when the permeability changes at a boundary. The B field is related to the H field by the equation $B = \mu_R \mu_0 H$, where μ_0 is the permeability of free space, and μ_R is the relative permeability. μ_0 is equal to $4\pi \times 10^{-7}$ T/A m (Section 1.13).

Capacitance. The ability of space to store electric field energy. In a capacitor it is the ratio of stored charge to voltage. The unit of capacitance is the farad (Section 1.7).

Capacitor. A circuit component designed to store electric field energy (Section 1.8).

Charge. A quantity of electrons gathered on a conductor or insulator. A coulomb of charge moving past a point per second is a current of one ampere (Section 1.2).

Coulomb. The unit of charge. One coulomb per second is one ampere.

D field. The electric field represented by charge distribution. The field lines are continuous across a dielectric boundary when there are no trapped electrons. The D field is proportional to the E field and is equal to $\varepsilon_0\varepsilon_R E$ (Section 1.9).

Dielectric. An insulating material such as glass epoxy (Section 1.6).

Dielectric constant (relative). The property of a material to reduce the electric field intensity (Section 1.7).

Dielectric constant (permittivity). The constant that relates forces between charges. The permittivity of free space ε_0 is 8.85×10^{-12} F/m (Section 1.3).

Displacement current. A changing electric field that is equivalent to a current and results in a magnetic field (Section 1.14).

Displacement field. The D field (Section 1.9).

E field. The force field that surrounds every charge. This is a vector field as it has intensity and direction at every point in space. Electric fields can attract or repel (Section 1.2).

Electric field. See E field.

Energy. The state of matter that allows work to be done.

Farad. The unit of capacitance. The ratio of charge to voltage (Section 1.7).

Faraday's law. The voltage induced in a conducting loop by a changing magnetic flux (Section 1.13).

Ferrite. A magnetic material in the form of a ceramic insulator.

Flux. Fields are often represented by lines through points of equal intensity. These lines as a group are often called flux. It is common to refer to magnetic flux. Lines that are closer together represent a more intense field (Section 1.13).

Force. That which changes the state of rest of matter.

Force field (Electric). A state that attracts, repels, or accelerates charges.

H field (static). The magnetic field associated with current flow. Magnetic fields are force fields that can attract or repel other magnetic fields. The H field is a vector field as it has intensity and direction at every point in space. The H field has units of amperes per meter (Section 1.12).

Henry. The unit of inductance. Abbreviated as H (Section 1.14).

Impedance. The opposition to current flow in a general circuit. For sine waves the real part is resistance, the imaginary part is reactance. Abbreviated as Z.

Inductance. The ability of a space in or out of a component to store magnetic field energy. The correct definition of inductance is the amount of flux generated per unit of current. The unit of inductance is the henry. Abbreviated L (Section 1.14).

Induction. The ability of a changing field to move charges on conductors or to create a potential difference in space (Section 1.13).

Induction field. The magnetic or B field.

Inductor. A component or conductor geometry that stores magnetic field energy (Section 1.15).

Lines. A representation of field shape and intensity.

Magnetic field. The force field surrounding current flow. The force field surrounding magnetized materials.

Magneto-motive force. The H field.

Mho. An earlier unit of conductance. It is now the siemen.

Mutual capacitance. Electric field coupling between different conductor geometries. Mutual capacitance is always negative (Section 1.10).

Mutual inductance. Magnetic field coupling between different conductor geometries. Mutual inductance can be of either polarity (Section 1.18).

Permeability. The ratio between the **B** and **H** magnetic field vectors. Abbreviated μ (Section 1.19).

Permeability of free space. A fundamental constant. The relationship between the induction or B field and the H field in space. Referred to as μ_0 (Section 1.13).

Permittivity of free space. A fundamental constant. The relationship between the D and E fields in a vacuum. Referred to as ε_0 (Section 1.9).

Potential differences. Voltage differences between conductors or between points in space.

Poynting's vector. A vector formed by cross multiplying the electric and magnetic field vectors. The vector is equal to the intensity of power flowing perpendicular through an increment of area in space. The vector can be zero behind many wave fronts (Section 1.20).

Reactance. The reciprocal of susceptance. The opposition to sinusoidal current flow in capacitors and inductors. In circuit theory the reactance of an inductor is a plus imaginary number of ohms. The reactance of a capacitor is a negative imaginary number of ohms.

Resistance. The opposition to current flow. The real part of impedance.

Self capacitance. The ratio of charge to voltage on the same conductor (Section 1.10).

Self inductance. The ratio of the magnetic flux to current flow in the same conductor geometry (Section 1.16).

Siemen. The unit of admittance. Reciprocal of impedance.

Tesla. The unit of magnetic induction (B field) (Section 1.13).

Transmission line. Parallel conductors that can direct the flow of energy.

Vector. A line (and arrow) in space where length represents magnitude, intensity, and direction of a force field.

Vector field. A parameter that varies both in intensity and direction in space. The most common fields are gravity, electric or E field, displacement or D field, magnetic or H field, and induction or B field.

Voltage. Correctly, a voltage difference. The work required to move a unit charge a distance in an electric field. Measurements are usually measured between conductors. Voltage at a point in space has no meaning (Section 1.3).

Work. Force times distance where the force direction is in line with the distance traveled (Section 1.3).

Answers to Problems

1. 1.73×10^{15} grains in a cubic foot.
2. $R = 0.274$ ohms.
3. Use admittances. 782 MHz.
4. 5 nH.
5. 250 mA.
6. 6.24×10^{17} electrons per second.
7. Avogadro's number divided by $62.55 = 9.5 \times 10^{21}$.
8. About 1 cm/second.
9. 50 mA.

2

Transmission Lines—Part 1

- Waves on transmission lines imply energy activity.
- They can indicate that electromagnetic energy is being transported.
- They can indicate that electric field energy is being deposited as magnetic energy.
- They can indicate that magnetic field energy is being deposited as electric field energy.
- They can indicate that electric field energy is being converted to magnetic field energy.
- They can indicate that magnetic field energy is being converted to electric field energy.
- Several of these activities can occur in parallel.

To understand which activity is occurring is not obvious. Measuring voltages is just a part of the story. The explanations that follow will take two full chapters.

2.1 Introduction

Chapter 1 introduced the electric and magnetic fields. These fundamental fields exist around static and moving charges. These fields are basic to all electrical activity including radiation. It was shown that it takes work to move fields into a volume of space. Without conductors, the fields we have been discussing simply radiate into space. To store and move energy in a circuit, fields must be confined in a conductor geometry that has capacitance and inductance. On a transmission line, energy is stored, converted, or moved by wave action. These transitions take place when sine waves are involved but it is difficult to present these mechanisms graphically. Step functions provide an opportunity to appreciate how energy is stored, moved, and converted.

There are many ways to move electromagnetic energy. Examples are lasers, fiber optics, light, radio transmitters, waveguides, and radar. Our interests are in moving waves on transmission lines on circuit boards. Even if other

Fast Circuit Boards: Energy Management, First Edition. Ralph Morrison.

methods are used to transport energy, in nearly every case the information being transported ends up being processed on a circuit board where there is access to data processing, data storage, and energy.

INSIGHT

When there are two or more conductors, energy can be transported on a controlled basis using electromagnetic fields at low frequency. This one fact makes it practical for us to power our homes and industrial society at 60 Hz. This same property allows us to run our logic circuitry where the spectrum can go to dc. It is this two-conductor transport of energy that is the focus of this book.

The role conductors play in confining fields is not entirely obvious. The parallel with a river is not perfect but it can help. Rivers do not leave their banks as it requires "going uphill." The water goes where the constraints permit flow. If there is a dam, the water rises. The dam has capacity measured in volume (acre-feet). The parallel with electrical phenomena is obvious but usually considered trivial. The river banks are the equivalent of the conductors. For a river that spreads out, the water will slow down. On a circuit board the spreading will allow energy to reflect and take alternate paths. Energy does not slow down. The slowing down takes place by storing parts of the energy in capacitance and inductance. Then later, the energy is released from storage by another wave action.

The problem in logic circuit board design is twofold. Energy must be stored near points of demand and then sent to receiving points to represent information. Both parts of the problem involve transmission lines and both parts are equally important. The time it takes to acquire energy is as important as the time it takes to transport this energy. As you will see, the available energy levels are quantized and limited by the geometry of the conductors. It is like a water pipe. Given a fixed pressure, the amount of water that can be delivered in a given time depends on the diameter of the conduit. Given a fixed voltage, the amount of initial current flow is defined by geometry of the conductors. As you will see, this is called characteristic impedance.

2.2 The Ideal World

We begin our discussion by connecting an ideal voltage source to parallel conductors called a transmission line. To keep the problem simple, the switch making the connection has no dimensions and closes in zero time. An ideal voltage source means that the source impedance is 0 ohm at all frequencies. This assumption is used in circuit theory to simplify the analysis. Fortunately, in logic, we can make these same assumptions and not get into trouble. This is because every connection to a voltage source or logic signal involves a transmission line and this added path dominates the flow of initial energy. Also, every practical logic switch has a finite operating time. There is both a time delay and a time to transition from open to its closing resistance.

A typical source of energy on a circuit board is a regulated power supply. It is common practice to use an active circuit for this function. A capacitor is usually placed across the output terminals of the supply to make certain that the voltage source is unconditionally stable for any load. First energy usually comes from this capacitor. In circuit theory, an ideal voltage source can both accept or provide current. The fact that an active source may not be able to accept current is of little concern. The fact that a transmission line is used to transport this energy is a more significant issue. A power supply for each active device is impractical. The approach we use involves decoupling capacitors which can allow current flow in either direction.

The partial differential equations that describe field behavior are difficult to use. Most of the equations we use to describe transmission lines can be developed by using a simple set of relationships. Logic signals, by their very nature, involve step functions of voltage or current. The leading edges of voltage have near constant slopes, which means that the time derivatives of voltage and current are nearly constant. This allows us to use simple algebra to develop the relationships that describe electrical behavior. Hopefully, this approach will provide insight and we will not need to solve differential equations.

2.3 Transmission Line Representations

The textbook representation of a transmission line usually shows increments of series inductance and shunt capacitance. For parallel traces, a circuit representation is shown in Figure 2.1.

It is convenient to use the language of circuit theory to get us started. In this network, the current that flows into the first increment of inductance charges the first increment of capacitance. As time progresses, a wave, observable as a

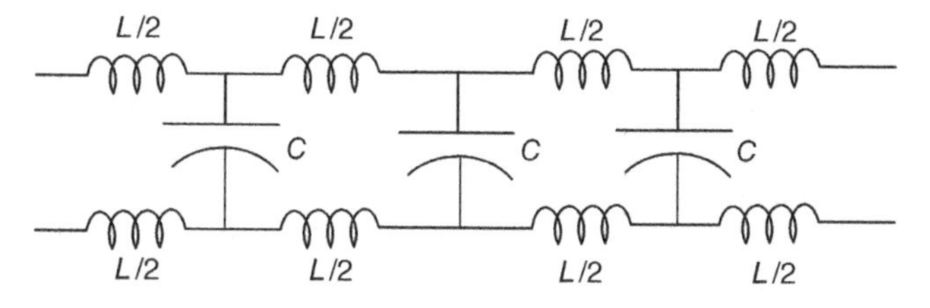

Figure 2.1 The lumped parameter model of a transmission line. *L* is inductance per unit length. *C* is capacitance per unit length.

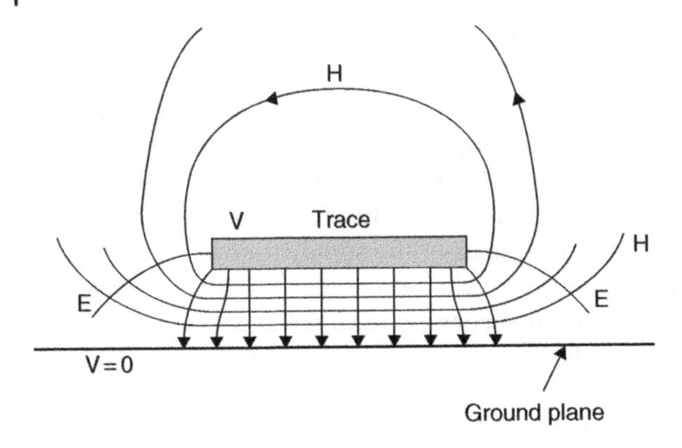

Figure 2.2 The field pattern around a trace over a ground plane (microstrip).

voltage, propagates through the network storing energy in distributed capacitance and inductance at a constant rate.

The electric and magnetic field patterns around a trace over a ground plane are shown in Figure 2.2. This field pattern propagates with time along the transmission line.

The voltage that is applied to the transmission line cannot be an ideal step function as this creates many philosophical problems including the need for infinite power. To avoid this issue, we assume that all voltages applied to a conductor geometry have a finite rise time. As you will see, what happens in this rise time is important and will be discussed in more detail later.

When a voltage is first connected to a transmission line, the resulting action is called a wave. This initial wave consists of a rising voltage and a rising current flow. The current is placing charge into the capacitance of the line. The voltage implies an electric field and the current flow implies a magnetic field. As the wave propagates, the current must continue to flow at a steady rate. Behind the wave front the electric and magnetic field intensities are unchanging. The magnetic field is supported by a forward and return current that must flow behind the wave front. This current crosses the transmission line at the leading edge of the wave. One way to describe this current flow is to say it is charging the capacitance of the line. Figure 2.3 shows the nature of the current as a wave moves along a transmission line.

We are very limited in our ability to make useful displays of wave action. It is difficult enough to display what takes place at one point on a transmission line, let alone observing the action as if it were a moving train. Waves do things that trains cannot do. Wave reflect and transmit at discontinuities and trains crash. This inability to see the waves contributes to our difficulties. We try to picture each event on a time scale we can deal with. We might be able to animate voltage but what is really happening is the movement of energy. Animating energy

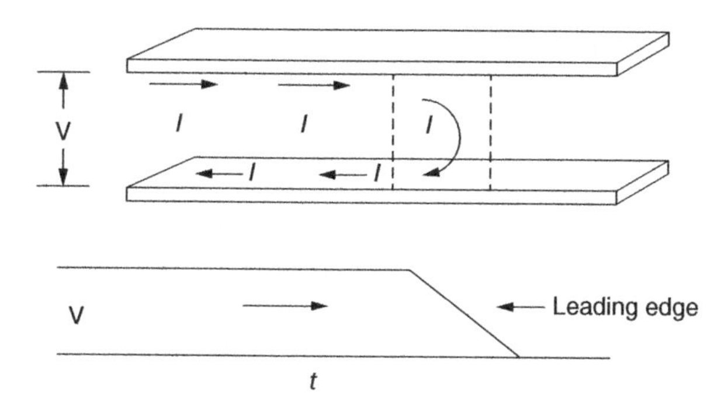

Figure 2.3 The flow of current in a wave as it moves along a transmission line.

or energy flow is not going to accomplish very much as this is really something we cannot see or measure. I will discuss this point later in the chapter. Energy flow is three dimensional and for traces it is cylindrical. Unlike water flow, the energy is confined even though the trace edges are not closed.

2.4 Characteristic Impedance

When a steady voltage V is first connected to a transmission line, current flows in the inductance of the line and charge is stored in the capacitance of the line. The charge Q in a unit of capacitance C is equal to the current I multiplied by time t. The capacitance C is thus equal to

$$C = \frac{Q}{\mathrm{V}} = \frac{It}{\mathrm{V}} \tag{2.1}$$

An increment of magnetic flux in space has an inductance equal to voltage multiplied by time t. The inductance L is the ratio of flux to current or

$$L = \frac{\varphi}{I} = \frac{\mathrm{V}t}{I}. \tag{2.2}$$

Dividing Equation 2.2 by Equation 2.1 and taking the square root, we see that

$$\left(\frac{L}{C}\right)^{1/2} = \frac{\mathrm{V}}{I} = Z_0 \tag{2.3}$$

where Z_0 is called the characteristic impedance of a transmission line. It has units of real ohms. It is a measure of the first current that will flow into a transmission line. Thus initially, a transmission line behaves like a resistance where the voltage wave shape is the same as the current wave shape. It also means that if the voltage is a sine wave the initial current is also a sine wave. Even though the characteristic impedance is measured in real ohms, there is no dissipation. The energy that enters the line is stored in the electric and magnetic fields on the line.

In circuit theory, an ideal transmission line has no losses and the input impedance for sine wave voltages is a reactance. This sinusoidal measure of a transmission line assumes all transient effects have attenuated. We will discuss sine waves and transmission lines later.

Circuit theory sees transmission lines as reactances. As strange as it sounds we are going to discuss transmission lines as resistances that store energy. This points out the difference in viewpoint between circuit theory and the one we take when we transport energy (logic). It is a different world.

The characteristic impedance of a transmission line on a logic circuit board is selected to optimize the performance of integrated circuits and at the same time limit cross-coupling. In many designs, 50 ohms is the preferred impedance level. Board manufacturers usually add traces that can be tested in the manufacturing cycle to verify that both plating and etching processes are held within expected limits.

In many designs, the trace lengths are short enough that the properties of the transmission line are not critical. The control of characteristic impedance is still important to control initial current demand, cross–coupling, and manufacturing repeatability. The control of characteristic impedance is a key factor in providing signal integrity (SI).

2.5 Waves and Wave Velocity

Waves are a common occurrence in nature. We see ocean waves and light waves and hear sound waves. We expect a flag or a tree to wave in a breeze. In many ways, we expect nature to use a wave mechanism to move or trap energy. We expect waves when potential and kinetic energy are linked together. The waves we are interested in move electrical energy back and forth in transmission lines. Within a short period of time, there may be hundreds of waves resulting from one switching event taking energy from a source. In one sense, there is only one block of energy in action but for many reasons it is practical to break up the action into waves that reflect and transmit at transition points. The rule we follow is that a wave at a transition point becomes one or two new waves and the first wave is discarded. At a 0 impedance point there is no transmission. At an open circuit there is no transmission but the arriving voltage doubles. The voltage at any point in a transmission is the sum of the wave amplitudes that pass that point. It makes no difference which direction the wave is traveling. This is why it is necessary to idealize a wave as being its transition (leading edge). A wave contributes a voltage to a point after the leading edge has passed that point. With the rules we have set up, a wave cannot pass a point more than once. We break up the action into waves that we can handle and then put the waves back together to see the end result which we can measure.

When a voltage is switched onto a transmission line, nature takes this as an opportunity to spread energy into a new volume of space. We know that energy begins to propagate as a wave along the line based on the geometry of the line. The leading edge adds charge to the line linearly with time. Using the definition of capacitance, we can relate the time it takes for the leading edge to pass one point

$$C = \frac{Q}{V} = \frac{It}{V} \quad \text{and} \quad C = \frac{\varepsilon_0 w \ell}{h} \tag{2.4}$$

where w, ℓ, and h are width, length, and conductor spacing, respectively. Solving for the permittivity of free space,

$$\varepsilon_0 = \frac{Ith}{Vw\ell}. \tag{2.5}$$

Using the definition of permeability and knowing the magnetic flux increases linearly during the passage of the leading edge, we can relate time to the dimensions of the transmission line.

$$\mu_0 = \frac{B}{H} = \frac{\varphi w}{\ell h I} = \frac{Vtw}{Ih\ell}. \tag{2.6}$$

The product of Equations 2.5 and 2.6 shows that

$$\mu_0 \varepsilon_0 = \frac{t^2}{\ell^2}. \tag{2.7}$$

INSIGHT

From Equation 2.7, the velocity ℓ/t of the voltage wave is

$$v = \frac{1}{(\mu_0 \varepsilon_0)^{1/2}}. \tag{2.8}$$

Using $\mu_0 = 4\pi \times 10^{-7}$ Tm/A and $\varepsilon_0 = 8.85 \times 10^{-12}$ F/m, the velocity is equal to

$$v = c = 300 \times 10^6 \text{ m/s} \tag{2.9}$$

which is the speed of light in a vacuum.

On a circuit board, traces are usually spaced from a conducting plane by a dielectric called glass epoxy. This material has a relative dielectric constant ε_R of about 4. At a GHz the value falls to 3.5. The permittivity of a dielectric is equal to $\varepsilon_R\varepsilon_0$. The wave velocity in a dielectric is the speed of light reduced by the square root of the relative dielectric constant or

$$v = \frac{c}{(\varepsilon_R)^{1/2}}. \tag{2.10}$$

The field energy of a typical surface trace on a circuit board is located mainly in the dielectric under the trace. The field energy above the trace is in air. The energy in air travels faster than the energy in the dielectric. The effect is to spread out the leading edge of the wave using the transmission line.

2.6 The Balance of Field Energies

The initial energy that enters a transmission line is simply VI times time. This must be equal to the sum of electric and magnetic field energy or

$$VI \times t = \frac{1}{2}LI^2 + \frac{1}{2}CV^2. \tag{2.11}$$

The ratio of electric to magnetic energy that enters a length of transmission line is

$$\frac{\frac{1}{2}CV^2}{\frac{1}{2}LI^2}. \tag{2.12}$$

Using Equation 2.3, it is obvious that this ratio is unity.

Thus, in every *initial* transmission, the moving energy is divided equally between electric and magnetic fields.

INSIGHT

The electric and magnetic fields that move energy are coupled to each other. At each point in space, the electric field vector is perpendicular to the magnetic field vector and both fields are perpendicular to the direction of energy motion. When energy is in motion, equal amounts of magnetic and electric field energy are required. As we see, in a transmission line, the two fields are static and at the same time they are the movement of energy. When viewed separately they are energy in storage. Transitions back and forth between storage and motion can only take place in the presence of conductors at the leading edge of a wave. In many situations there can be energy storage and energy motion in the same space.

It is important to note that the magnetic and electric fields contribute equally to any losses.

2.7 A Few Comments on Transmission Lines

The transmission lines used on a typical printed circuit board are traces over a conducting plane or traces between two conducting planes. The literature treats several other trace arrangements including side-by-side traces, asymmetrical traces between conducting planes, and traces between grounded trace pairs. Beside trace arrangements, transmission lines can be coaxial cables or ribbon cables. There are issues of cost and application but our concern is basically SI on circuit boards. The important factors we consider include terminations, reflections, crosstalk, energy containment and energy movement. The specific trace geometry will be mentioned only when it is relevant.

Circuit theory stresses loop or node analysis. For transmission lines, the return current path forms a loop that has little to do with circuit analysis. If the loop area relates to cross-coupling or radiation we will be very specific. The return current path is critical because it defines the characteristic impedance of the transmission line. In our discussions, a controlled return current path is assumed even though it is not shown.

We often use single-line diagrams showing individual waves traveling in either direction. The waves will be shown as having a leading edge that is symbolic. In practice the leading edge may spread out over the entire transmission path making an actual wave depiction somewhat impractical. Whatever the picture, if energy is in motion it will move at the speed of light in the dielectric involved. The symbolic leading edge of a wave indicates the direction of wave travel which is of course key information. The leading edge also makes it easier to see how the voltage derivative explains current flow in the capacitance of the line.

INSIGHT

Every conductor pair is a transmission line. Conductors include the earth, conduit, building steel, racks, cabinetry, power conductors, as well as traces over or between conducting planes on a circuit board. Given an opportunity, electromagnetic fields will couple to every conductor pair because it is a way to "spread out" the energy. In particular, all conductor paths have a lower characteristic impedance than free space. In the design of a logic board the object in design is to control the field paths so that logic/energy will arrive where needed in an effective way using a dedicated path.

2.8 The Propagation of a Wave on a Transmission Line

To start, let us connect a voltage source to a transmission line. An initial voltage would require immediate field energy. Since this is impossible, an initial voltage must start from 0 and rise in a finite time. Since there cannot be magnetic field energy present in zero time, the current must also start from 0. In a rise time, the full voltage and full current must be present at the input terminals. In the second rise time period the wave has progressed at the speed of light and the voltage and current are at their limiting value at the input terminals. The ratio of voltage to current flow is the characteristic impedance of the line as given in Equation 2.3. Along the transmission line, where the voltage is in transition, a displacement current crosses the transmission line in the changing D field. This current path is continuous from the voltage source, along the trace, through the space between the conductors, and on the return conductor. In the third rise time period the current loop is one increment longer and a fixed voltage and current are present on the line. Stated again, the forward current crosses the line in the space of the leading edge as it moves down the line. The changing D field in this space is equal to the return current flow. This is a very important idea to grasp.[1]

INSIGHT

Behind the wave front, the E and H fields are unchanging. Poynting's vector indicates that energy is in motion but there is no such thing as a Poynting's meter. The only motion that can be detected are the slow moving charges on the conductor surfaces. This is one area where circuit theory gets it right. Current times voltage equals energy flow per second or power. We measure these parameters on conductors but the energy flows in space.

A mechanical parallel may help in explaining energy flow. Consider the drive shaft in an automobile. Energy is coupled to the wheels by a drive shaft, yet no part of the shaft is moving forward. The energy is transferred in the shaft by rotation and in shaft torque. The energy is carried forward in rotational stress in the steel. The stress is greatest near the outer surface of the shaft. The velocity of a pressure wave in steel is surprising. It is nearly 10,000 mph. The parallel with the transport of electromagnetic field energy is food for thought. This leads to the conclusion that electromagnetic fields are a stress in space that transfers energy at the speed of light. In a transmission line the fields and the conductors play a

---

1 One of Maxwell's equations is in part $\partial H_Z/\partial y - \partial H_Y/\partial z = i_Z + \partial D_X/\partial t$. This general equation relates the pattern of the magnetic field H in the xy plane to the time changing D field in the x direction and to the current flow I_Z. Recall that $D = \varepsilon E$ where E is the electric field in volts per meter and ε is the permittivity of the dielectric. This equation shows that both a current or a changing D field creates a magnetic field. In transmission lines there is no current flow in the space between conductors except at the leading edge. There is current flow in a resistor termination and on the surface of the transmission line conductors.

The second and third part of this equation (not shown) relates the H field in the xy and xz directions to the current flow in the z and y directions.

It is convenient to depict waves in motion using short rise times. In some logic, the transitions in voltage occur over an entire trace length and wave action is not easily displayed. Displacement currents still flow but across the entire line at the same time.

role in propagating a stress in space. Wheels, gravity, and the mass of a chassis are needed or the drive shaft could not be stressed. Conductors play this role in a transmission line. Without conductors, space cannot be electrically stressed at dc.

It is difficult to accept the idea that the fields are not moving although they are changing at leading edges. The problem is that at dc, the fields are unchanging and it is the energy that is moving at the speed of light. In a logic transmission, the fields behind the leading edge are static and they are carrying energy. To say the fields are moving is a common misconception. It serves no purpose to make this assumption. The only motion we can sense is at the wave front where we can make measurements. The idea of field motion is somewhat supported when we modulate the energy flow sinusoidally. This modulation allows us to see the wave movement but in no way can this activity be assigned to actual energy motion. Waves indicate transition.

The movement of energy in a transmission line is often paralleled with the action of a group of pendulums. When the first pendulum hits a series of suspended balls, the last pendulum receives the energy and moves. The velocity of the energy motion depends on the elastic properties of the balls. This parallels the transfer of energy in a drive shaft where mechanical stress is propagated. In this analogy, the energy flow is an impulse.

2.9 Initial Wave Action

The wave action that first propagates on the transmission line carries energy in an electric and magnetic field. As the wave propagates, energy is deposited in the capacitance and inductance of the line. This deposition occurs as the leading edge passes each point. The energy behind the wave front is carried at the speed of light in both the E and H fields although the fields are unchanging. When the wave reaches the end of the line, energy has been deposited in the line capacitance and inductance and field energy is still moving along the entire line toward the termination (load).

INSIGHT

The fields behind the first wave action serve two purposes. First it represents the energy that has been stored in the line and at the same time it is involved in the movement of energy behind the wave front. The fields behind the wave front are unchanging and yet energy is in motion. The fields thus have the ability to store and move energy at the same time. In the drive shaft analogy, the torque on the shaft stores static energy, the rotation of mass stores kinetic energy, and this energy does not move down the shaft to the wheels.

If the transmission line is terminated in a resistor equal to the characteristic impedance Z_0, the energy in the wave flows along the line and into the resistor and all apparent wave action stops. There is field energy in motion and energy stored in the line capacitance and inductance. The resistor will dissipate energy until the voltage source is removed and all the energy in the transmission line reaches the resistor. This energy cannot be put back into the power source. This wave action is shown in Figure 2.4.

A second representation showing wave action is given in Figure 2.5.

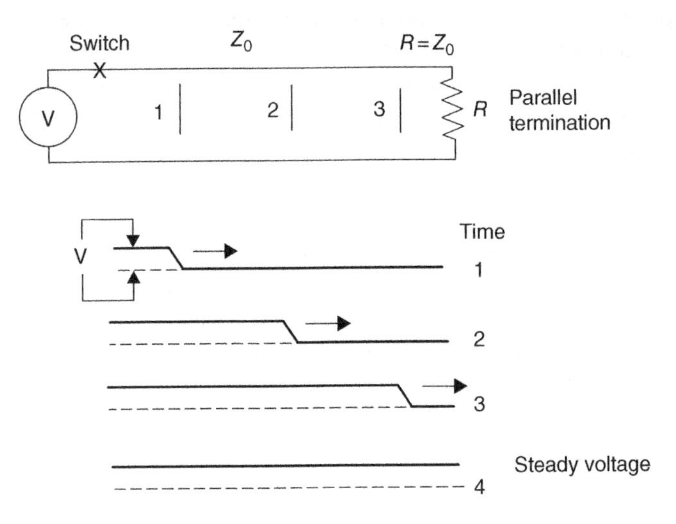

Figure 2.4 The wave action associated with a transmission line shunt terminated in its characteristic impedance.

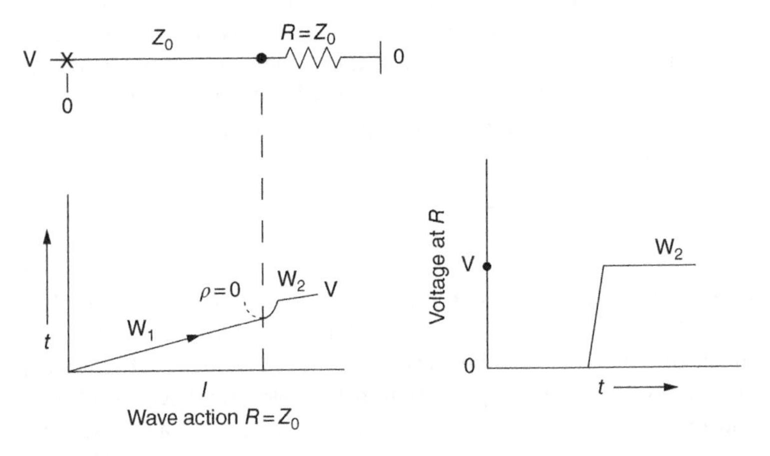

Figure 2.5 The wave action on a transmission line shunt terminated in its characteristic impedance.

In this representation, time is shown on the vertical axis and wave activity along the transmission path is shown on the horizontal axis. The U-shaped "cups" on the diagram represent points in time and position where waves will transition. Usually a transition is a reflection back and a transmission forward. For the cup at time $t = 0$ and position ℓ, the closure of the switch results in a transmission W_1 forward into the line. When the wave reaches the resistor R, there is no reflection. The forward transmission wave W_2 is a voltage V.

In some applications, it may be desirable to terminate a line in a Zener diode in series with a terminating resistor. The Zener clamps the reflected voltage and terminates the line until wave action stops. This avoids damage to the input gate and allows the final voltage to hold without dissipating energy.

2.10 Reflections and Transmissions at Impedance Transitions

There are many places on a circuit board where the characteristic impedance of the transmission path changes. This can include sources of logic and energy, as well as at logic terminations. The wiring from an IC to pads or from pads to a capacitor are examples of short transmission lines that are often not considered. A termination can be an open circuit, a resistor, or another short transmission line. Terminations involving a general RLC network will not be considered.

On many board designs, the leading edges may transition slowly enough so that reflections do not occur. We assume zero rise times to develop equations for reflection and transmission coefficients. The criteria for when to apply these equations is given in Section 2.13.

At a transmission, the initial forward voltage is V_1, the reflected voltage is V_R, and the transmitted voltage is V_L. Using circuit theory the relationship between these three voltages is

$$V_1 + V_R = V_L. \tag{2.13}$$

Assume the initial transmission line has a characteristic impedance of Z_0 and the terminating transmission line or resistance is Z_L. The currents must satisfy the relationship

$$\frac{V_1}{Z_0} - \frac{V_R}{Z_0} = \frac{V_L}{Z_L}. \tag{2.14}$$

Eliminating V_L between Equations 2.13 and 2.14 and solving for the ratio V_R/V_1

$$\rho = \frac{V_R}{V_1} = \frac{Z_L - Z_0}{Z_L + Z_0} \tag{2.15}$$

where ρ is called the reflection coefficient.

The reflected wave has a voltage equal to

$$V_R = V_1 \ \rho \tag{2.16}$$

Eliminating V_R between Equations 2.13 and 2.14 and solving for the ratio V_L/V_1,

$$\tau = \frac{V_L}{V_1} = \frac{2Z_L}{Z_L + Z_0} \tag{2.17}$$

where τ is called the transmission coefficient.

The transmitted or forward wave is equal to

$$V_L = V_1 \tau \tag{2.18}$$

The reflection coefficient ρ at an open circuit is 1. This means that the reflected wave amplitude equals the forward wave amplitude. The resulting voltage at the open circuit is double the forward wave voltage. Energy continues to flow into the transmission line doubling the voltage. If the transmission line is terminated in a short circuit then the reflection coefficient is −1. This means that the voltage of the reflected wave is reversed in polarity. Short circuits do not dissipate energy; they simply reflect the wave. If the termination is equal to Z_0, then there is no reflection and the energy flows directly into the termination.

The transmission coefficient τ at an open circuit is 2. The voltage at the output terminals is double the arriving wave voltage. If the output is shorted, the transmission coefficient is 0. If the output is terminated in Z_0 the transmission coefficient is 1, which means that there is no reflection and all the energy proceeds forward.

When a wave reaches the end of a transmission line terminated in a voltage V, the terminating impedance is 0 and there is no transmission. The reflection coefficient is −1 and the reflected voltage is the terminating voltage less the reflected voltage. In effect, the reflected wave returns the voltage on the line to V.

The reflection and transmission equations given above are very general and apply to any waveform. For sine waves, the voltages and loads can use the complex notation of circuit theory. See Appendix A for a review of this notation. When sine waves are involved, sums and differences are also sine waves. When other waveforms are involved this simplification is not available. The voltages on the following transmission line terminations are all step functions. Each reflected or transmitted wave is assumed to be another step function with the same rise and fall time. Each combination of waveforms is unique. This makes it more difficult to represent each method of termination. This is why it is helpful to use two different representations to display the wave action.

2.11 The Unterminated (Open) Transmission Line

The voltage representation of this wave action is shown in Figure 2.6. The rise time is assumed to be a small part of the forward transit time.

A second representation of this wave action is shown in Figure 2.7. Note the reflections at the voltage source and at the open circuit. At the open circuit, two new waves W_2 and W_3 are generated. At the voltage source there are no transmitted waves. The reflected waves are the incident waves reversed in polarity.

When the reflected wave W_3 reaches the voltage source, a second reflection W_4 takes place. This reflected wave W_4 must have an amplitude of $-W_3$.

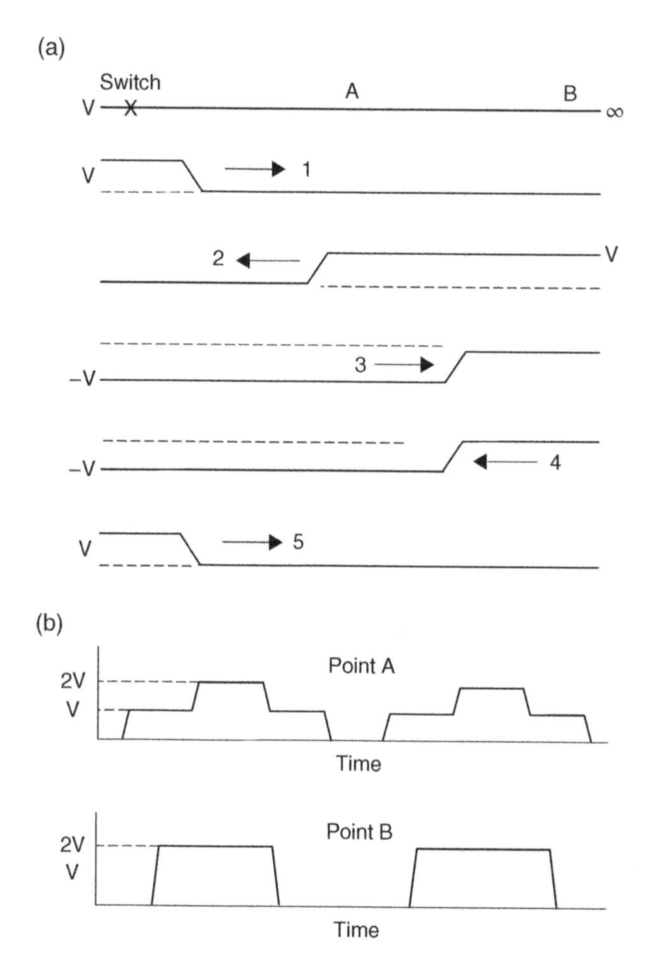

Figure 2.6 The voltage waveforms on an ideal open circuit transmission line for a step input voltage. (a) Individual waves and (b) The sum of the waves.

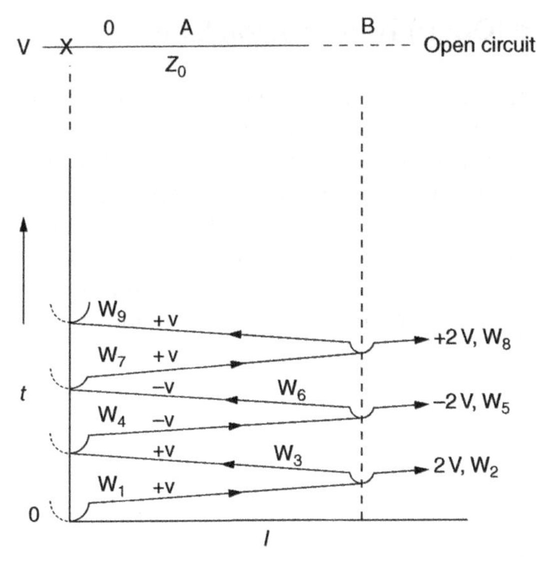

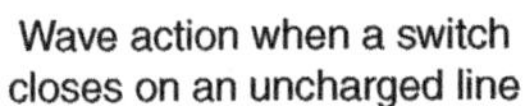

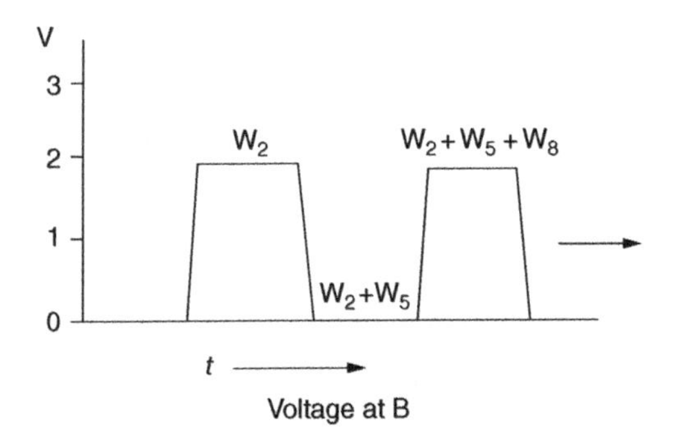

Figure 2.7 The voltage waveforms on an ideal open circuit transmission line for a step input voltage.

The current for this wave is supplied by half the charge stored in the line capacitance.[2] The conversion to a negative current and a magnetic field takes place along the line at the leading edge of W_4. The voltage at the source supplies no further energy to the line.

---

2 The third wave in the unterminated line converts electric field energy to magnetic field energy. One of Maxwell's equations is in part $\partial E_Y/\partial x + \partial E_X/\partial y = \partial B_Z/\partial t$. This equation relates the pattern of the electric field E in the *xy* direction to the time rate of change of the magnetic field B in the *z* direction. Recall that $B = \mu_0 H$ where μ_0 is the permeability of free space. The letter subscripts indicate coordinate directions. The voltage that is produced by the changing magnetic field at the leading edge is approximately the E field intensity times the distance between conductors in the transmission line. The second and third parts of this equation (not shown) relates the E field in the *yz* direction to the changing B field in the *x* direction and the E field in the *xz* direction to the changing B field in the *y* direction.

When W_4 reaches the open end of the line, the negative current cannot flow in the open circuit; so, a new wave must return the output voltage to 0. The leading edge of this wave W_6 converts the remaining energy in the capacitance to current and stores energy in the magnetic field of the line. The full cycle is complete when the last or sixth wave reaches the source. Without losses, this wave sequence will repeat over and over.

Four waves (W_3, W_4, W_5, and W_6) constitute one full cycle of what might be called a resonance. Two waves place electric field energy on the line. Then the next two waves convert this energy from an electric field to a magnetic field. After two round trips of waves, the next electric field energy comes from the existing stored magnetic field, not from the voltage source. This assumes no losses.

The open transmission line behaves just like the LC tank circuit of circuit theory. The difference is that step voltages are involved instead of sine waves. The lesson is simple. Without a way to lose energy, an ideal open transmission line must oscillate. Note that the voltage pattern at points A and B are different.

INSIGHT

Consider a voltage V connected to an unterminated line. When the switch closes, the first wave of current is $I = V/Z_0$. When the first energy reaches the end of the line, the energy cannot go off into space. The arriving energy continues to flow placing charge into the capacitance of the line. This requires a reflected voltage or second wave that moves toward the source at the speed of light. At this point, *the reflected wave and the energy are moving in opposite directions*. This contradicts the idea that each wave has a trailing voltage and current. This doubling of charge deposited on the line means that the voltage is being doubled along the entire line. Note that we can theoretically monitor the movement of voltage but not the movement of energy. Note that Poynting's vector behind the first reflected wave is 0 as there is no current flow and thus no magnetic field. The conversion of energy is not the flow of energy.

The wave action and the flow of energy are both moving at the speed of light. This implies a relative velocity of $2c$ which would seem to violate the rules of relativity. What is observed depends on the position and velocity of the observer. The apparent velocity does not involve moving matter and can be greater than c.

There have been observations in astronomy where the apparent velocity of an event appears greater than the speed of light. A parallel with ocean waves will help explain this phenomenon. When an ocean wave hits a sea wall at a slight angle, the return wave appears to move along the wall at a very high velocity. If the angle is 10° and the forward wave travels at 5 mph, the wave peak will seem to move along the wall at about 50 mph. This has been called a phase velocity. It does not represent the velocity of the energy flow.

In the oscillating transmission line, the voltage source supplies no energy after the first round trip. The waves that are moving back and forth convert field energy back and forth between the magnetic and electric forms. In these waves, there is no net energy flow, only energy conversion. Remember, we showed that the initial wave carried exactly equal amounts of electric and magnetic field energy.

INSIGHT

The reflection process described above takes place along a transmission line. Energy is not reflected, only transferred back and forth between capacitance and inductance. When fields in free space reflect from a conducting surface, energy is not constrained by a conductor geometry; so there is no energy storage.

The fact that an ideal open transmission line oscillates when excited by a step wave should not be a surprise. There is inductance, capacitance, energy, and no intentional losses. In circuit theory, this is a tank circuit that supports a decaying sine wave. In this example the reflections are step functions. We know that in practice there are losses and the voltage soon stabilizes at the level of the initial wave. When the logic state is changed back from a 1 to a 0, the voltage changes but the energy stored on the capacitance of the line is not removed. This means that the line must oscillate. In circuit theory voltages just disappear from the circuit. The truth is that when energy is transported or is in transition it must eventually be dissipated and this takes time.

When the logic switch shorts the open line, the leading edge of a wave converts the energy in the line capacitance to magnetic field energy which we can sense as current flow. This current does not come through the logic switch as there is no new energy added to the line. When this wave reaches the open end of the line, the reflection converts the magnetic field energy back to electric field energy which appears as a negative voltage. What we see is wave action that appears as a square wave of voltage that goes from +V to −V. It may seem strange, but when there is voltage there is no current and when there is current there is no voltage. As before, when there is energy on a transmission line and there is no terminating resistor, the energy will transfer back and forth in a pattern we recognize as a resonance.

The presence of a negative voltage assumes the logic gate will accept this voltage without causing a diode to conduct. A conducting diode dissipates energy and this modifies the wave action. There is always the issue of whether this dissipation is within the rating of the integrated circuit. A parallel external diode would only provide protection if it conducted at a lower voltage.

A germanium diode conducts at a lower voltage than silicon. It is temperature limited and may not be an acceptable protection device.

The following sections discuss various transmission line configurations. The wave action is discussed mainly in terms of voltage transitions. The function of the individual waves is not always discussed. Note that some energy is dissipated when energy is moved in and out of a magnetic field. Some energy is dissipated when it is moved into and out of a capacitance.

2.12 The Short-Circuited Transmission Line

The pattern of voltage waves is shown in Figure 2.8a and the corresponding pattern of current waves is shown in Figure 2.8b. Unlike the open-circuit case, there is no oscillation and the current builds up in a step manner.

The initial voltage wave stores energy in both the line capacitance and inductance. The first reflected wave converts electric field energy to magnetic field energy. This results in doubling the current flow behind the wave front. When this wave returns to the source, the voltage source must support three units of forward current: the initial current, the converted current, and the current for the next forward wave. After the next round trip of wave action, the current increases to 5 units. The arrows in the figures indicate the wave direction and not the current direction which always points toward the short on the transmission line. This is another example of where the current flow direction and the wave directions are not the same.

A second representation of wave action for a shorted transmission line is shown in Figure 2.9.

2.13 Voltage Doubling and Rise Time

In Section 2.10, we saw that there is voltage doubling on an open transmission line. This occurred because we assumed zero rise time for the initial step voltage. With a finite rise time, a doubling of voltage can only occur if the line is longer than 2.5 times the distance traveled by the wave in the rise time. In many circuit board applications, line terminations are not required. Because clock rates are increasing, this problem needs to be addressed as each design is considered.

At 1 GHz, a wave travels about 15 cm in one clock cycle. The distance traveled in a rise time is about 1.5 cm. Voltage doubling will begin to appear if the line is longer than 3.7 cm. Since there can be about 1.5 cm of line inside a component, the maximum trace length before voltage

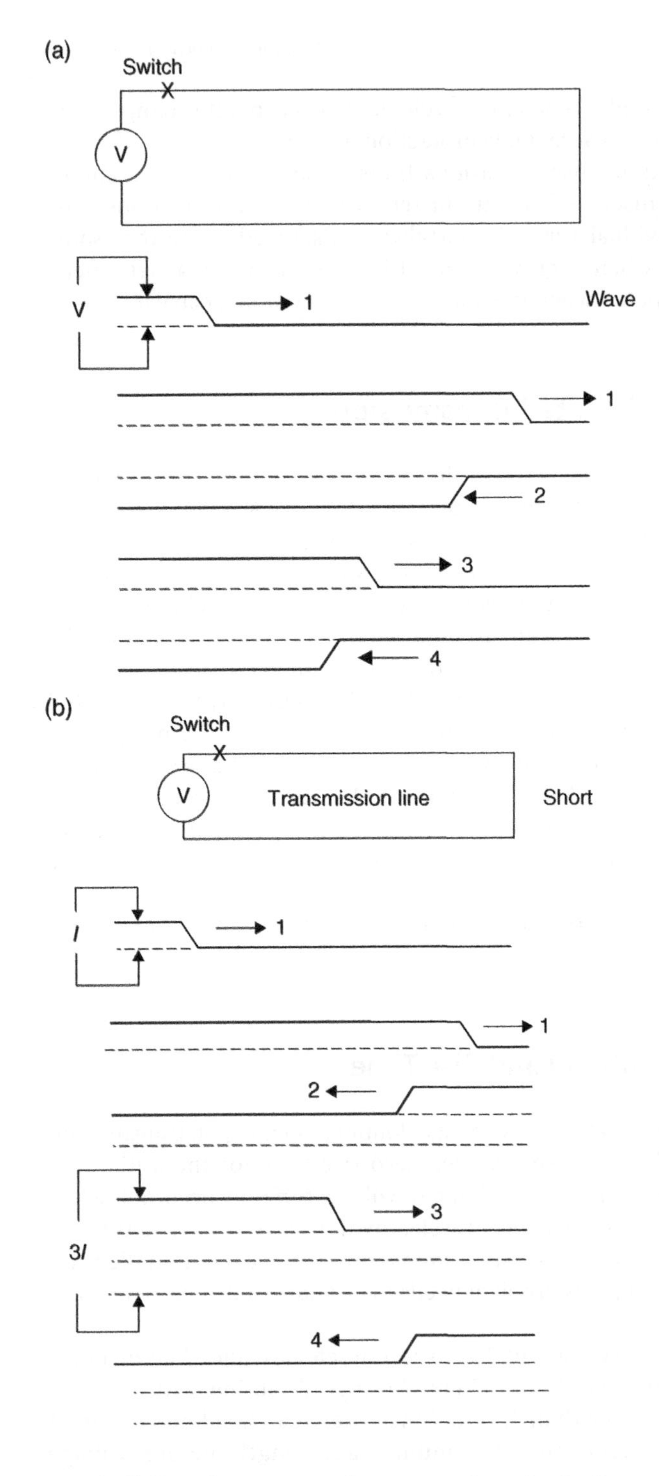

Figure 2.8 The voltage pattern of waves on a short-circuited transmission line. (a) Individual waves and (b) The sum of the waves.

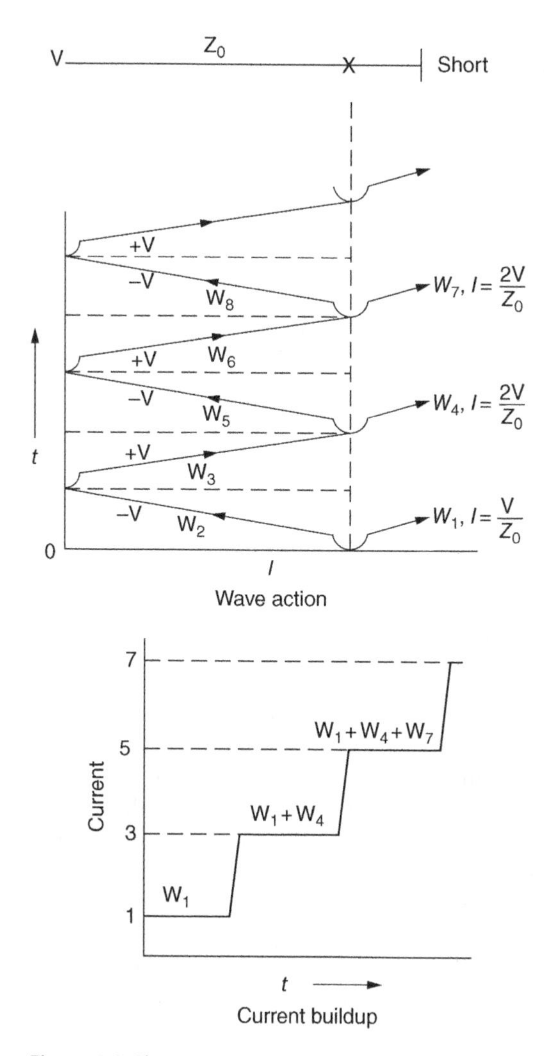

Figure 2.9 The staircase current pattern for a shorted transmission line.

doubling occurs is about 0.7 cm. It is safe to say that termination resistors are needed above 1 GHz. If the forward wave is at half voltage, the reflection will double the voltage and no termination resistors are needed (see Section 2.16).

This doubling action is shown in Figure 2.10 where the line length is given in terms of the distance traveled by the wave. This figure can easily be scaled by increasing the line length and using this same factor on all of the rise times.

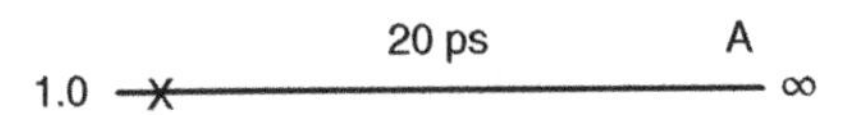

Signals are measured at point A.
A 1-V signal is at the source.

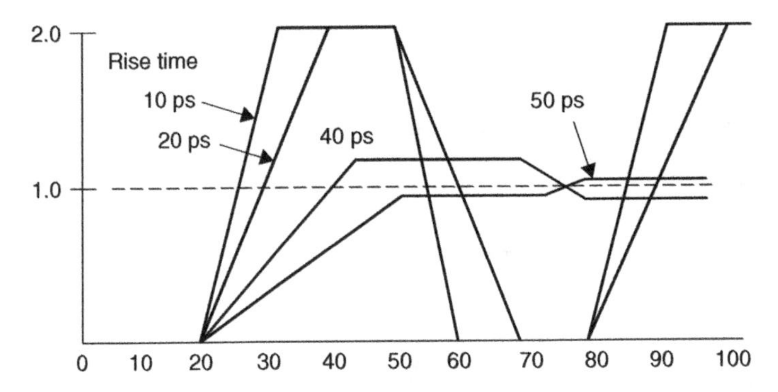

Figure 2.10 Rise times and the reflections from an unterminated transmission line.

2.14 Matched Shunt Terminated Transmission Lines

In Figure 2.11 an ideal voltage source V is connected to a length of 50-ohm transmission line terminated in a matched (shunt) 50-ohm resistor. When the switch X connects the transmission line to the resistor, the voltage at the switch drops to 2.5 V and a wave of −2.5 V travels back toward the source. At the source, this wave reflects and a wave of +2.5 V travels forward. At the load this voltage adds to the initial voltage making the total voltage 5 V. There is no reflection as the line impedance matches the termination resistor and all wave action stops.

This wave action occurs any time a circuit makes a request for energy from a remote point. The intervening transmission line determines the level of initial current and the time it takes for the request to be met. The voltage source could be a decoupling capacitor.

This method of line termination dissipates energy as long as the logic switch is closed. It is used on long lines where voltage doubling is a problem. When the switch is opened, a wave must travel toward the voltage source that sets the current to 0. In an ideal line the energy in motion before the switch opens cannot be lost. In this case, the arriving energy is stored in the line capacitance. This means that the return wave doubles the voltage across the open switch which can destroy the switch. If the logic is to be disconnected from the transmission line it should be done on the other side of any termination. This avoids any voltage doubling. Once the switch opens, the line has inductance, capacitance, and energy and there is no terminating resistor. The result is an oscillation on a resonant transmission line.

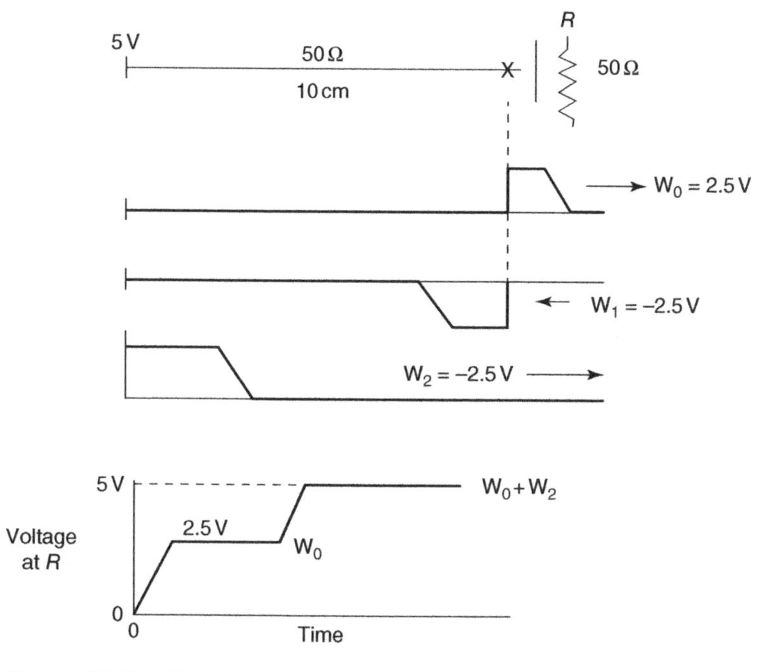

Figure 2.11 Matching shunt termination using a remote switch.

A second method of displaying wave action on this transmission line termination is shown in Figure 2.12.

When the switch closes, a wave W_1 of amplitude V/2 enters the resistor. Wave W_2 moves toward the voltage source and reflects. There is no wave transmission into a voltage source. The reflected wave W_3 reverses the polarity of W_2. When W_3 reaches the resistor R there is no reflection as the resistor matches the characteristic impedance of the line. The transmission coefficient is unity and the wave $W_4 = W_3$ enters the resistor. Two waves of amplitude V/2 have arrived at the resistor R making the final voltage equal to V.

If the transmission line characteristic impedance is higher than the load, then the reflected waves will be smaller and more wave action is required to bring energy to the load. This is shown in Figure 2.13.

If the resistor R is 25 ohms in Figure 2.11, then the wave into this resistor would be 0.666 V and the reflected wave would be −0.333 V. The current for this wave comes from the charge stored in the line capacitance. At the source, the reflection reverses the wave polarity and a wave of +0.333 V moves forward. At the load, another reflection and transmission takes place. The voltage at the load increases in the steps 0.666 V, 0.888 V, 0.962 V, 0.986 V, and so on. In most practical situations, it is not possible to see these step voltages.

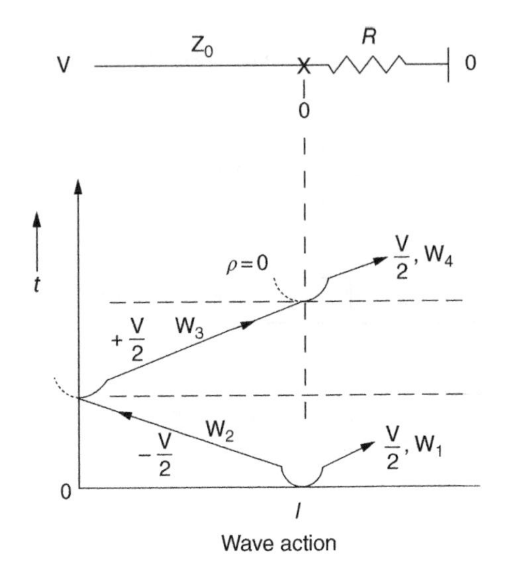

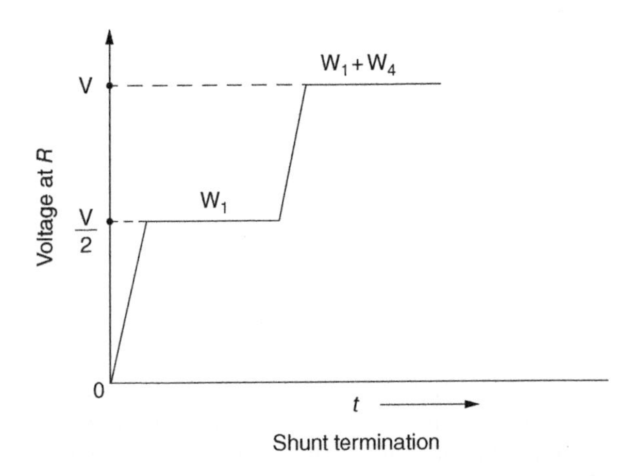

Figure 2.12 A matching termination using a remote switch.

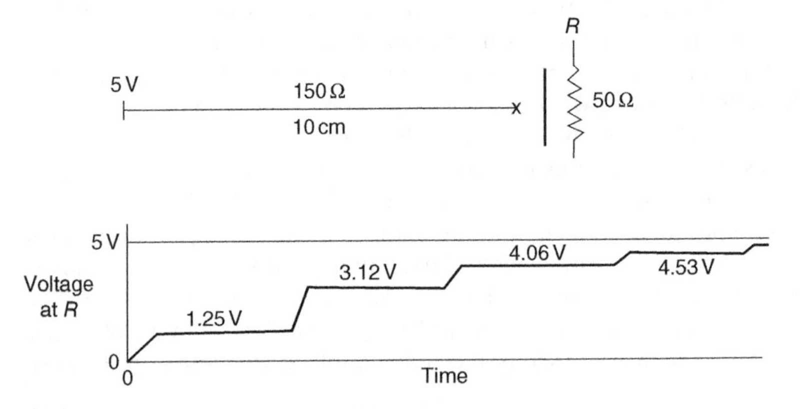

Figure 2.13 The voltage at a termination when there is a mismatch in impedances.

What is seen is a smooth leading edge that appears exponential in character. Remember that wave character involving transmission lines is not the same as transient behavior described by linear circuit theory. In circuit theory, this exponential rise in voltage would be explained by assuming a parasitic series inductance or a parasitic shunt capacitance. In this example, the time it takes for the voltage to reach its final value involves the path length as well as the capacitance, the dielectric constant, and the inductance in the transmission path. Assuming the delay is the result of inductance alone is incorrect. It is the result of moving energy which is a very different picture. This wave action is shown in Figure 2.14.

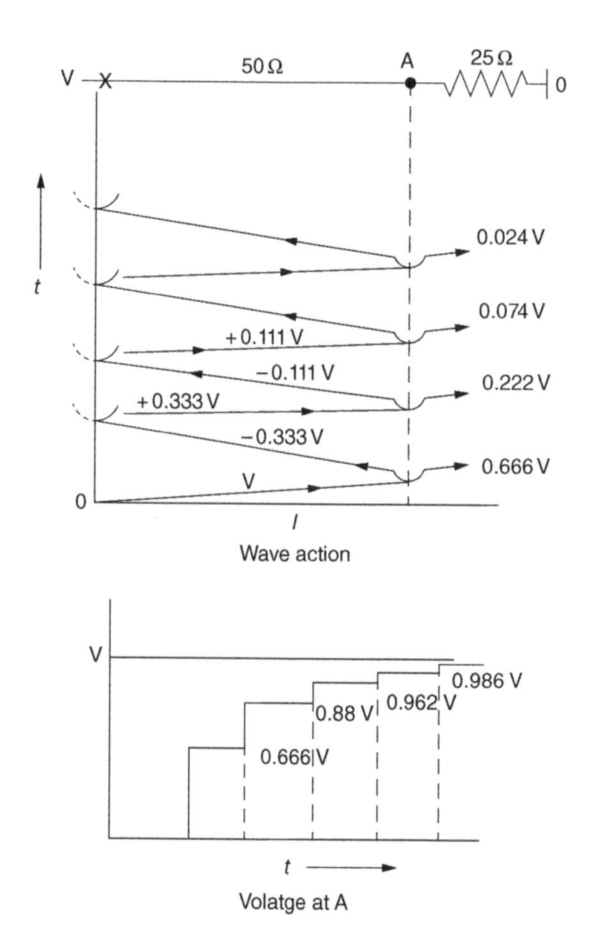

Figure 2.14 The voltage at a termination when there is a mismatch in terminating impedance.

2.15 Matched Series Terminated Transmission Lines

In Figure 2.15 an ideal voltage source is connected to a series switch, a line matching series resistor of 50 ohms, and a length of 50-ohm line that is unterminated. When the switch connects a voltage to the line, a wave of amplitude V/2 travels down the line. At the open end of the line the voltage doubles to V. The reflected wave converts the magnetic field energy to electric field energy. When this wave reaches the matching resistor at the source, the voltage across the resistor is 0 and all wave action stops. This is an ideal transmission line configuration for short rise time logic. The matching series resistor must be located very close to the voltage source and the logic switch for this wave action to take place. If the integrated circuit is built with an internal series line-matching resistor, then no external series matching resistor is needed. When the switch returns the logic voltage to 0, the forward wave is –V/2. At the open circuit the wave voltage is doubled and the resulting voltage at the open circuit is 0 V.

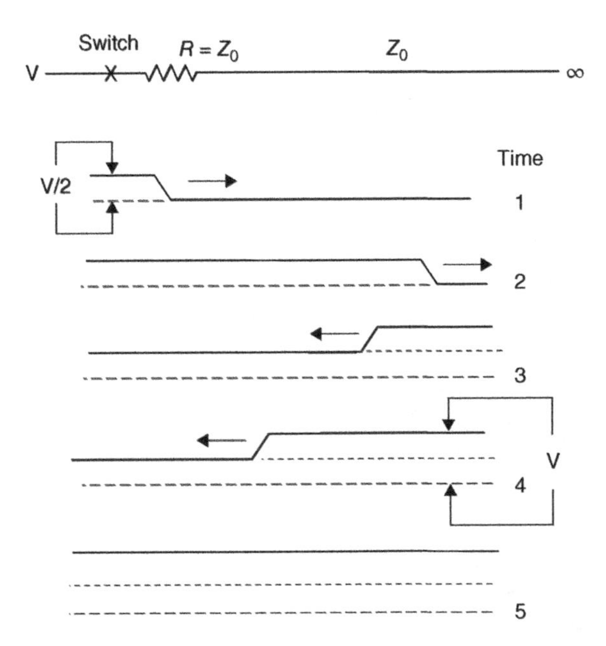

Figure 2.15 A series (source) terminated transmission line.

2.16 Extending a Transmission Line

Logic operations can involve extending transmission lines from a source of energy such as a decoupling capacitor through a switch to a remote gate G. In Section 3.2, we will discuss the fact that there are several sources of immediate energy in the IC. Assume for the moment that these energy sources are not present. The transmission line to the left of a switch is labeled d_1. The line to the right is labeled d_2 and stores no energy. This transmission line arrangement is shown in Figure 2.16. Initially the switch is open.

At the moment where d_1 connects to d_2, waves must propagate on each transmission line. The connected voltage drops to V_1. The voltage of the wave going left is $-(V-V_1)$. The voltage of the wave going right is V_1. The energy in d_1 in a distance *s* must travel a distance 2*s* to get to the point *s* on d_2. The initial energy in a distance *s* is $\frac{1}{2}CV^2$. The energy in motion is half magnetic. Therefore, in a distance 2*s*, the energy in the electric field is $\frac{1}{4}CV^2$. This equation shows that $V_1 = V/2$. This means that there is a wave growing in length traveling in both directions moving energy from d_1 to d_2. The wave moving left is depleting energy in d_1 and this energy is moving right into d_2. If the relative dielectric constant is 4, the wave velocities are $c/2$. A full understanding of what occurs requires a relativistic treatment.

The energy flowing into d_2 comes from d_1. Since the wave in d_1 is traveling left and the energy in d_2 moving right, the rise time for the forward wave in d_2 is doubled over the rise time in the initial switching action. This doubling of rise time does not usually occur as there is usually enough local energy in parasitic capacitance to fill in the gap. The wave that travels right into d_2 carries energy forward. This energy flow as required is half electric and half magnetic. When the wave front traveling left reaches the voltage source, the wave reflects and a positive wave moves forward. This wave converts the magnetic field energy moving left into electric field energy that is stationary (stored in the line capacitance.) This wave action returns the energy to d_1. Remember that the combined stationary E and H field along the line implies energy flow.

When the wave moving forward in d_2 reaches the open end of the line, the wave reflects. The wave front converts arriving magnetic energy into electric field energy doubling the voltage on the line. This wave action places energy in the line capacitance of d_2. When this wave reaches the forward wave, all wave activity stops as the currents are 0 and the voltages are equal. Wave action has moved a block of energy from d_1 to d_2 and supplied a block of energy to d_1.

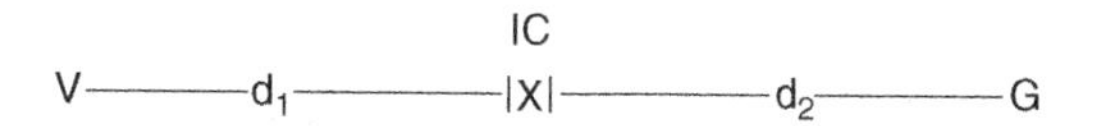

Figure 2.16 A typical transmission line.

At first glance this seems like a perfect solution where no terminating resistor is needed. In practice, several logic elements can make a demand for energy at the same time. If these demands are made on one transmission line to an energy source the wave action does not stop as described above. A separate power trace for each logic line is impractical. The only viable solution is to provide an energy source in or at the IC. In this situation, the full voltage moves forward and voltage doubling can do damage. This is the reason why a series matching resistor is needed.

INSIGHT

Wave action that converts electric field to magnetic field poses a philosophical problem. We are used to thinking of electric field energy in a capacitance as being static. An increment of electric field energy is not a problem. Thinking of an increment of magnetic field associated with an increment of current as being static is not as obvious, yet that is an important part of understanding wave action on transmission lines. The leading edge of a wave converts an increment of electric field energy into an increment of magnetic field which implies an increment of current. Having an increment of static current out in the middle of a transmission line is counter intuitive and yet that is what happens.

The movement of energy requires both an electric field and a magnetic field which are static. These fields are the energy in motion which we cannot see or measure directly. This phenomenon can only occur when there are conductors. I am reminded of the flashlight where nothing changes but energy moves. There is a static electric and magnetic field.

2.17 Skin Effect

A full treatment of skin effect on circuit traces is complex. An approximate picture of this effect can be helpful.

The textbook approach to skin effect assumes a plane electromagnetic field reflecting from an infinite conducting plane. The field is sinusoidal so that the surface currents that flow are also sinusoidal. The field intensity at any depth is a function of frequency, conductivity, and permeability. For copper the relative permeability is unity. A skin depth at a given frequency is the depth where the field intensity is reduced by a factor of 1/e or 36.8%. The frequency that is used when logic is involved is obviously a compromise. The frequency that is often used is $1/\pi\tau_R$ where τ_R is the rise time of the logic.

The attenuation factor for an infinite conducting plane and a plane sinusoidal wave is

$$A = e^{-\alpha h} \tag{2.19}$$

where h is the depth and

$$\alpha = \left(\pi \mu_0 \sigma f\right)^{\frac{1}{2}}. \quad (2.20)$$

The permeability μ_0 of space is $4\pi \times 10^{-7}$ T m/A. The conductivity of copper is $\sigma = 0.580 \times 10^8$ A/V m and f is frequency.

When $h = 1/\alpha$ the field intensity is reduced by a factor of 36.8%. This is called one skin depth. For copper at 1 MHz, the skin depth is 0.65 μm. At 100 MHz, the depth is 0.065 μm. This approximation is a reminder of just how little of the copper gets used in a single transmission. If the copper is plated this is where the initial current flows.

It is interesting to point out that a static voltage without current flow along a transmission line requires a surface charge. These charges are the first to move when wave action takes place. Just a reminder, electron velocity in a transmission line is only centimeters per second.

INSIGHT

Transmission line processes are not represented by circuit theory. When a voltage is connected to a transmission line, the source can only react to the characteristic impedance. Until a first reflected wave returns, the source cannot react further. This is not the case in circuit theory where the source senses the entire network in the first moment. We deny this idea of instantaneous reaction and yet we do not fault circuit theory. The way of the future requires some changes in approach because time delay is a real issue that must be faced in moving energy.

Problem Set

1. If the dielectric constant increases by 2%, what is the change in characteristic impedance?
2. A transmission line is 10 cm long. The dielectric constant is 4 at 10 MHz and 3.5 at 1 GHz. What is the arrival time difference for sine waves?
3. Show that $1 - \tau = \rho$
4. A 5-V wave on a 50-ohm line terminates on a 60-ohm line. What is the voltage of the transmitted wave? What is the reflected voltage?
5. A 50-ohm line terminates in two 50-ohm lines. What are the reflected and transmitted voltages if the initial voltage is 5 V?
6. What is the field intensity at two skin depths?

7 An unterminated transmission line is oscillating. The dielectric constant is 3.5. What is the frequency if the length of the line is 5 cm?

8 A request for energy is made over a 50-ohm line that is 4 cm long. The dielectric constant is 3.5. The logic is 5 cm away. Assume two round trips are needed. What is the delay?

Glossary

Characteristic impedance The ratio of initial voltage to initial current on a transmission line. On an ideal line the ratio is constant until a reflected wave returns to the source (Section 2.4).

Conductivity A measure of a material to conduct current. It is the reciprocal of resistivity. The resistance of a conductor is $R = \rho l/A$ where ρ is resistivity, l is the length of the conductor, and A is the cross-sectional area (Section 2.17).

Exponential Growth or decay of physical phenomena where the activity is proportional to the amount present involves a natural constant e which is 2.718. The voltage on a capacitor discharged by a resistor is said to be exponential decay (Section 2.14).

Flux Usually a representation of the magnetic B field. Line density indicates field intensity.

Free space Ideally a vacuum removed from conductors or charges. Practically, the space around a circuit not near conductors.

Ground plane A conducting surface usually larger than any associated components. On a circuit board, it can be a conducting layer (Section 2.3).

Leading edge The voltage pattern at the front of a wave.

Line A transmission line. A trace running parallel to a conductor.

Lumped parameter A circuit representation of a conductor geometry where distributed parameters are represented by series and parallel elements. On a transmission line, every increment of the line has inductance and capacitance. The lumped parameter model is a series of inductors interspersed with shunt capacitors (Section 2.3).

Matching termination Usually a resistor equal to the characteristic impedance of a transmission line. A terminating transmission line with a matching characteristic impedance.

Reflection coefficient The ratio of a reflected wave amplitude to a forward wave amplitude at a transmission line termination. For a transmission line this ratio can cover the range −1 to +1. At a short circuit the reflection coefficient is −1. For an open circuit the reflection coefficient is +1. This means the voltage is doubled (Section 2.10).

Relative permittivity The dielectric constant.

Resonance A condition where energy is in constant transition between two storage mechanisms. On a transmission line the energy transfers back and forth between electric and magnetic fields. On a pendulum, energy transfers back and forth between energy of motion and energy of position (Section 1.11).

Rise time The length of time it takes a wave to transition from 10 to 90% of final value. It is usually a voltage measure. Other definitions are frequently used.

Series termination A termination, usually a resistor, that is in series with the source voltage on a transmission line.

Shunt termination A termination that is placed across a transmission line usually at the end of the line.

Skin depth A measure of how far into a conductor surface currents penetrate. One skin depth implies the current falls to 1/e or 36.8% of the surface value.

Skin effect A phenomena that occurs in conductors where changing currents tend to flow on the surface of conductors. The measure is made for sine waves but applies to any waveform (Section 2.17).

Step voltage A voltage that changes value in zero time. This is of course impossible as the energy stored in parasitic capacitance would have to appear in zero time and this would require infinite power. In practice a step voltage has a finite rise time (Section 2.11).

Termination The impedance at the ends of a transmission line. A termination can be an open circuit, a short circuit, or a resistor.

Transmission coefficient The ratio between a transmitted wave and an arriving wave at a transmission line termination. The coefficient can cover the range 0–2. At a short circuit, the coefficient is 0. At an open circuit, the coefficient is 2, which means the voltage is doubled (Section 2.10).

Transmission line A conductor geometry that supports the flow of electromagnetic energy.

Unterminated line A transmission line terminated at an open circuit. The actual termination is a small capacitance that can modify the leading edge of an arriving wave.

Wave A transition in voltage that travels along a transmission line usually characterized by a leading edge followed by a steady voltage. A wave can carry, deposit, or transform energy. The term wave is also used for sinusoidal transmissions where there is no leading edge. Waves can be confined by conductors or they can move in free space. There are wave patterns that appear stationary. A steady pattern moving in or on a media can also be called a wave.

Answers to Problems

1 The capacitance increases by 2% and the characteristic impedance drops by 1.414%.
2 42 ps.
4 $\tau = 1.091$, $\rho = 0.091$.
5 Both the reflected and transmitted waves are 3.333 V.
6 One skin depth is 36.8%. The square of this percentage is 13.5%.
7 802 MHz or four different wave patterns.
8 The distance travelled is 27 cm at a velocity of 160 m/μs. The time is 1.7 ns.

3

Transmission Lines—Part 2

3.1 Introduction

There is no polarity associated with field energy. One field does not cancel another. Consider the wakes of two boats on a lake. The waves will cross each other without any cancellation. Currents and voltages add and subtract in circuit theory and this represents one of the basic differences in how electrical phenomena is viewed. Another problem is that a rise in electric field intensity might occur while the magnetic field intensity is decreasing. This means that the term "rise time" must be used carefully. It would have been better to refer to transition time instead of rise time but that would defy convention. In the discussions that follow, it will be convenient to stress waveforms in terms of voltage. Keep in mind that there will usually be an associated current waveform present. These waveforms are drawn two-dimensionally and what they represent are three-dimensional fields. For traces on a circuit board the fields are usually cylindrical. When the path spreads out there are reflections and delays. This spreading of waves is three dimensional and not easily shown on the printed page. When there is spreading, there can be problems. The reader is asked to use his or her imagination to picture the wave action as outlined by the current path. The fact that the field lines are perpendicular to the direction of wave motion adds to the difficulty.

3.2 Energy Sources

Circuit boards can be operated from batteries, utility power, or even solar cells. In many cases the energy is supplied by an external power supply and brought to the circuit board on a cable or through a board-edge connection. There is often an on-board voltage regulator that conditions the voltage to the requirements of the circuitry. Converters are available that can supply different voltages. A shunt capacitor is often placed across active voltage sources to ensure

Fast Circuit Boards: Energy Management, First Edition. Ralph Morrison.

stability and supply peaks of current if needed. The energy source we concentrate on is the simple decoupling capacitor. This capacitor is usually supplied energy over a transmission line from a nearby power supply.

The voltage that is switched onto transmission lines implies a supply of energy at the point of switching and this is the energy that is discussed in this section. In the analog world, wiring of a few inches to a regulated voltage would be entirely adequate. In logic, the problem is much more complex. It is important to accept the fact that initial current is controlled by the characteristic impedance of any connecting transmission lines. Fortunately, there are several immediate sources of energy near points of demand. This energy is stored in the capacitance of "connected" transmission lines.

A logic switch is usually a part of the die in the integrated circuit. In the die there may be other logic lines connected to the power potential through logic switches. Some of these lines extend through wiring to pads and out on to the board. Each of these connected paths is a transmission line storing electric field energy. In fact, many of these lines are physically no different than the lines that are usually labeled "power." Energy taken from a connected logic trace introduces interference on that line. Nature pays no attention to the labels we attach to conductors. We, on the other hand, have to look beyond labels to see what is happening.

A second source of energy can be the capacitance between the power and ground planes if they are used. Drawing from this source of energy adds interference to logic using this same space. The third source of energy are any decoupling capacitors mounted on the board near the integrated circuit. If these local energy sources were not available, then a logic switch would connect two transmission lines together. One line might lead to a voltage source and one is a logic line. At this midpoint, the voltage would drop by about 50%. Because there are these local sources of energy, any drop in voltage is moderated. These sources of energy are shown in Figure 3.1.

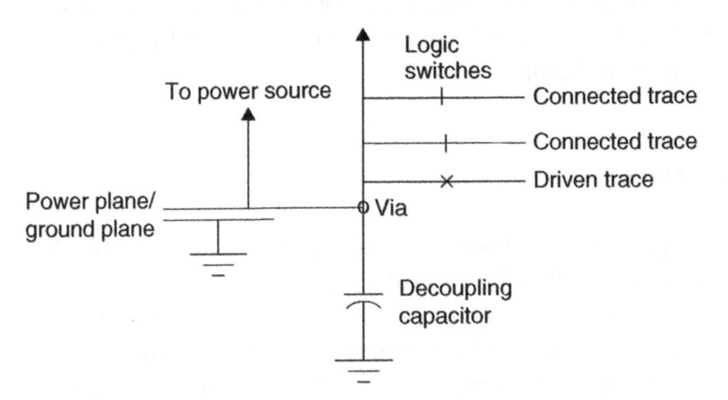

Figure 3.1 The first energy sources when a logic trace is connected to the nearest power conductor.

3.3 The Ground Plane/Power Plane as an Energy Source

A limited amount of embedded energy is available from a ground/power plane pair. The capacitance is $\varepsilon_R\varepsilon_0 A/h$, where A is the area and h is the spacing. The energy that flows into a central point from parallel conducting planes is pulled from concentric circles or rings. The characteristic impedance of a ring falls off inversely proportional to the radius. To increase the energy storage, a closer conductor spacing plus a higher dielectric constant is possible. The wave velocity falls off as the square root of the dielectric constant and this slows down access to this energy. The best way to increase the energy storage without adding delay is to decrease the conductor spacing. The characteristic impedance at the entry point will still be approximately 50 ohms. A single decoupling capacitor can outperform this energy source and provide a 3-ohm source impedance. The economics are obvious. Accept any energy that is available but rely on decoupling capacitors as the principle energy source.

A power plane using the entire board is not recommended as traces using the ground/power space will couple to noise. Reducing the spacing between conducting planes makes it impractical to control the characteristic impedance of traces using this same space. At first glance, it seems like a good idea to make double use of this space. The problem is that space for logic transmission should be dedicated and the space used for energy storage must be shared.

3.4 What Is a Capacitor?

INSIGHT

A capacitor is a component that stores electric field energy. It is a single-port device which means it cannot supply energy and be recharged at the same time. Capacitors are available with various entry configurations. Regardless of how the component is designed, it is usually mounted as a stub on a transmission line and this alone limits its performance. The water analogy on a hydroelectric dam makes the point. Water is never supplied and taken from the same conduit. If you want more water on demand you provide a larger conduit. The size of the lake does not change the flow rate. This is also true for capacitors. The initial flow is not controlled by capacitance but by the characteristic impedance of three paths: the path requesting energy, the path in the capacitor, and the path to the termination.

Even if capacitors could be made using two ports, the problem is that a high dielectric constant slows the flow of energy from the second port to the point of demand. This points out that the problem is much more complex than just capacitor geometry.

The simplest capacitors are formed by placing a dielectric between two conductors. The capacitance is a function of conductor area, dielectric thickness, and the dielectric constant. Small size is achieved by using thin dielectrics with high dielectric constants. For example, BST (Bismuth Strontium Titanate) can have dielectric constants as high as 12,000. The capacitance can vary with voltage and temperature, but these variations are not considered important in decoupling applications where the exact amount of stored energy is not important.

Capacitors are basic to all circuit design. They serve to block dc, to shape frequency responses, to reject noise, and to supply energy. On a board dedicated to processing logic, decoupling capacitors are the principle source of immediate energy.

INSIGHT

An important factor that is often not considered is that field energy in a dielectric moves slower than it does in air. If the dielectric constant is 10,000, then energy travels at 1% the speed of light. This means that a capacitor that is 0.1 cm long appears to be electrically 10 cm long which implies a capacitor acts like an unterminated transmission line stub.[1] Wave action that moves stored energy from a stub must make many round trips on both the stub and on the connecting transmission line.

We have already discussed the fact that it takes both inductance and capacitance to move energy along a transmission line. A capacitor has a series inductance that can be measured by observing its natural frequency. In circuit theory, this inductance might be considered a parasitic element. In logic, this inductance plus capacitance makes up a transmission line that can move energy. In circuit theory, the reactance of a stub depends on line length, frequency, the dielectric constant.[2] In logic, the movement of energy from a capacitor is not defined by its stub parameters (reactances) but by the entire transmission path from the capacitor to the point of energy demand. The traces that connect the capacitor to a load are an impedance mismatch the flow of energy from the capacitor requires considerable wave action. The characteristic impedance of

1 Bill Herndon, working at Motorola in Phoenix Arizona, in the early 1960s proposed that capacitors are transmission lines. This is not a new idea.

2 The "sinusoidal response" of a transmission line or stub usually means the ratio of current flow to voltage at one frequency and all transient effects are ignored. In logic, the opposite is true. The transient behavior of transmission lines is important and the steady state behavior is assumed to be unimportant. The ratio of first voltage to first current is not the same as the steady state ratio.

a capacitor at its terminals is about 3 ohms. If the rise time is long enough then wave action will not be apparent. Delays in moving energy are usually blamed on series inductance which is an over simplification.

In applications where a group of transmission lines carry parallel bit streams, the demand for energy can be high. Increasing the capacitance of a single decoupling capacitor will not usually support the step demands for more energy. A solution that is effective is to cluster a group of small capacitors at the integrated circuit. These parallel sources can provide the initial energy that no other conductor geometry can supply. An obvious help is to reduce the characteristic impedance of the line that connects the decoupling capacitors to the IC. If the characteristic impedance of the trace is reduced from 50 to 25 ohms, it will double the energy provided by each wave. The best solution is to mount capacitors at the IC pad and limit the transmission path to the die/pad distance.

INSIGHT

Step waves move energy in a transmission line in electric and magnetic fields. In a sinusoidal analysis where reactances are involved, the energy flow involves phase relationships between voltages and currents. Mixing disciplines leads to confusion. A capacitor has all the physical attributes of an unterminated transmission line, namely, two conductors separated by a dielectric. A capacitor is a transmission line with no access to its open end. Treating it as a transmission line allows all the paths of energy flow on a circuit board to be treated the same way.

3.5 Turning Corners

There is a sense that there is "mass and velocity" associated with any energy flow. This is a common feeling with no foundation. In Section 2.7, the point was made that the electric and magnetic fields behind a wave front are static and yet the coupled fields are carrying energy at the speed of light.

Many efforts have been made to detect reflections when a transmission line turns a right angle. Assumptions have been made that the traces should follow an arc or at least the sharp edges should be chamfered. There is no evidence that the wave action is affected by these efforts. This understanding must be extended to other minor transitions in transmission. It also applies to via transitions through board layers provided that the return current path is carefully controlled. Right-angle transitions are not the problem. The spreading of fields is far more important. Another way to look at this type of transition is to notice the difference in path length as energy turns the corner. For a 5-mil trace width the path length difference at the edges of the trace is only 5 mils. This is one wavelength at 2300 GHz. It is not surprising that we cannot see a reflection at a right-angle bend.

3.6 Practical Transmissions

The intent in logic is to move energy from points of storage onto transmission lines. These transmission lines terminate on logic gates and the voltages at these gates will then control an action in an integrated circuit at a next clock time.

As we have shown, the first energy for a transmission comes from several sources (see Figure 3.1). The next level of complication relates to transitions in the characteristic impedance as waves progress back from the die to the voltage source and forward on a trace to a second pad and on IC wiring to a nearby logic gate. The voltage source in the following figure is ideal, but in practice it could be a decoupling capacitor with a 3-ohm characteristic impedance.

In this figure, there are about 100 waves shown, most of them in the short wiring lengths in the ICs. These waves travel in the space between a conductor and an unspecified return path. The vertical axis is time (approximately). Note that the wave velocities vary depending on the dielectric constant and the transit time depends on the path length. The voltage at the open gate is the sum of all the waves that exit to the right on the diagram. The arrival time depends on the path taken. Obviously, the wave action continues beyond the 100 waves shown with ever-decreasing wave amplitudes.

In Figure 3.1, it takes wave action to move energy inside an IC. This means that the total wave action is even more complex than the wave action shown in Figure 3.2. If all the transmission segments were 50 ohms, then there would be far fewer transitions. This example shows that higher bit rates will require more attention to detail both in IC packaging and on the board pad and trace structure. The reason is simple. All of this wave activity results in a delay in moving information. The energy is available but it is delayed as a result of wave action.

INSIGHT

A delay in the arrival of a logic signal can be a problem. It is sometimes blamed on a single parameter such as an inductance or a capacitance. In reality, a delay is often the result of hundreds of reflections and transmissions. Faster logic will not function unless these transitions are considered. It only works when an arriving wave converts energy to an electric field and no other wave arrives.

On a circuit board, the wave actions described above rarely lasts more than a 100 ns. For GHz logic, picoseconds are counted and every aspect of delay is important. A reminder: There are three orders of magnitude between 100 ns and 100 ps.

N.B.

The delay caused by wave action involves the transitions located in the IC between the die and the circuit board pads. A part of the problem involves the use of shared commons to limit the pin count. Shared commons are not necessarily bad as this is what happens for traces over a conducting plane. What is important is that signal spaces be kept separate and this implies that the return currents use dedicated paths on the same conductor. There are many ways to configure a single return conductor so the fields use a different space. As an example, a single round conductor with multipole slots could support say eight separate conductors forming eight transmission lines. I did not say this was easy to do. I mention this because making an IC faster does not mean this speed can be used. Speed is a system problem that requires the entire transmission path for each signal be controlled. The first change in approach is to provide energy storage inside the IC. This energy can be replaced at a slower pace from outside. The second change is to control the characteristic impedance of each fast logic line that connects the IC to the pad structure. After this, it becomes the designer's problem to complete the transmission paths on the board.

In Figure 3.2, a switch connected a logic line to 0 V and opened the connection to the supply voltage. The energy stored on the line begins to move. The wave action that follows must eventually dissipate the stored energy in switch and trace resistances. Some of this energy will be radiated. None of this energy can be returned to the source. This shows again that energy flow is a one-way "downhill" path.

3.7 Radiation and Transmission Lines

Radiation in the form of a pulse can occur in nature. Examples are lightning, ESD, and solar flares. Step waves are the type of signals that radiate from transmission lines. Electrical engineering has methods of describing any electrical event in terms of its spectrum. Measurements are made by scanning the arriving field energy using narrow-band filters. In the logic systems we consider, the signals have a frequency content that is determined by rise times and clock rates. Repetitive signals are characterized by signals at the fundamental clock rate and its harmonics. A single step function is characterized by a continuous spectrum with an upper frequency characterized as having a frequency of $1/\pi\tau_R$, where τ_R is the rise time. See the comments made in Appendix B.

At 1 MHz, a wavelength in space is 300 m. Unless a significant effort is made, circuits do not radiate very much energy at this low a frequency. A 100-MHz square wave has frequency content well above 500 MHz and radiation efficiency

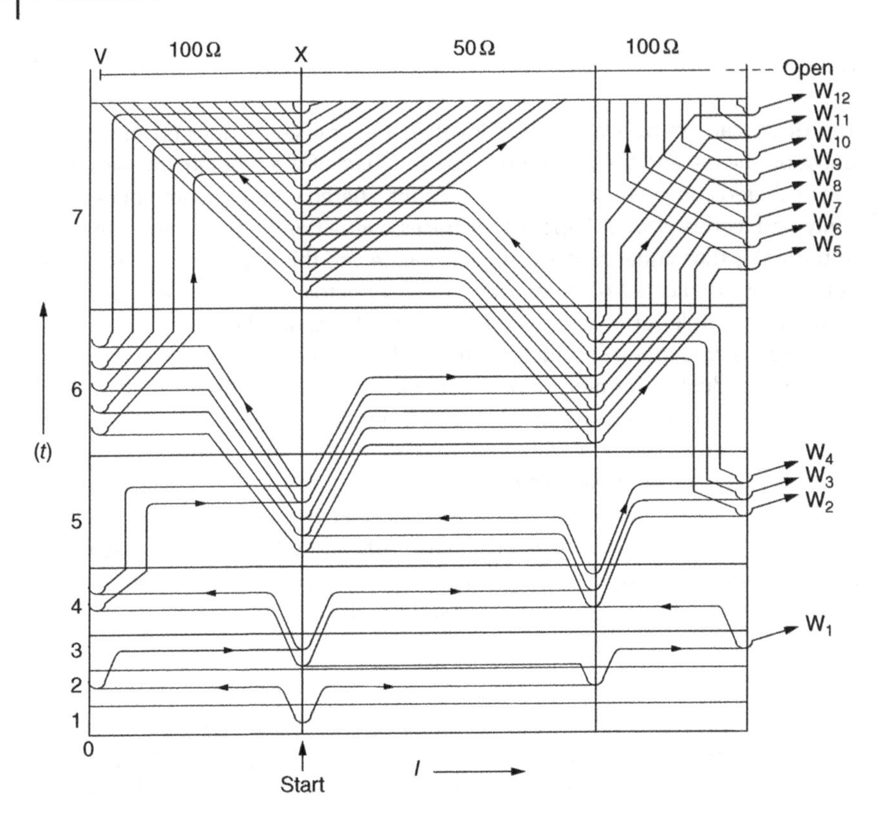

Figure 3.2 Wave action for a simple logic transmission. Note: Each wave is on a different time scale.

increases as the square of frequency. This means that at this clock rate, the spectrum will allow for radiation. Errors in layout that invite board radiation are discussed in Chapter 5.

Many products in common use today function based on receiving radiation in the range 1 MHz to 10 GHz. Uncontrolled radiation could cause real problems for these products. For this reason, permitted levels of radiation are controlled by government regulations. A cell phone is expected to radiate but only over a controlled portion of the spectrum. Circuit boards intended for logic operations must pass strict radiation tests before they can be placed on the open market. These rules are important as the number of circuits in operation that radiate or receive radiation at any one time in an urban area can number in the thousands. The count is growing exponentially. The many circuit boards operating in a modern automobile is an example of the problem.

Coaxial transmission lines are used to confine fields. On a circuit board, transmission lines are not coaxial and some fields extend above surface traces and along trace edges. Regulations do allow for some radiation. Cables that enter a board can bring in field energy or provide a path for energy to leave the board.

Radiation is discussed in more detail in Chapter 5.

> **INSIGHT**
>
> Radiation from a circuit board transmission line only occurs at the leading edge of a wave. In effect, the leading edge is a moving transmitting antenna. Of course, if the leading edge is spread out over the transmission path then the entire path radiates. The energy moving behind the leading edge requires a static electric and magnetic field that does not radiate. Remember that wave direction and energy flow direction can be opposites.

3.8 Multilayer Circuit Boards

Component density and pin counts on circuit boards have risen steadily. This, in turn, has resulted in the need for multilayer boards to accommodate the large number of interconnecting traces, most of which are transmission lines. Trace widths have dropped to accommodate larger trace counts and also to reduce board size. This need has reduced the diameter of drilled holes and made it more difficult to solder plate inner surfaces.

Board manufacturers can supply multilayer boards with layer counts over 60.

The manufacturing process involves plating and etching of copper layers separated by insulators. Many of the layers are laminates (cores) that are drilled and plated before being bonded together in a stack. Bonding a stack usually involves layers of partially cured epoxy called prepreg. Bonding of a layup takes place under pressure at an elevated temperature. A four-layer board is made up of a core placed between sheets of prepreg and copper. Figure 3.3 shows a four-layer board layup.

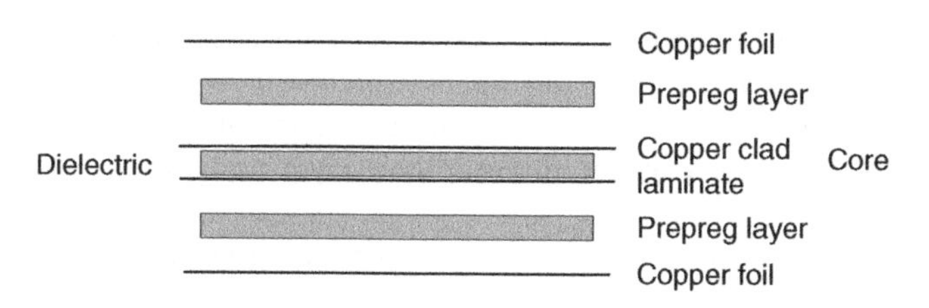

Figure 3.3 A four-layer board layup.

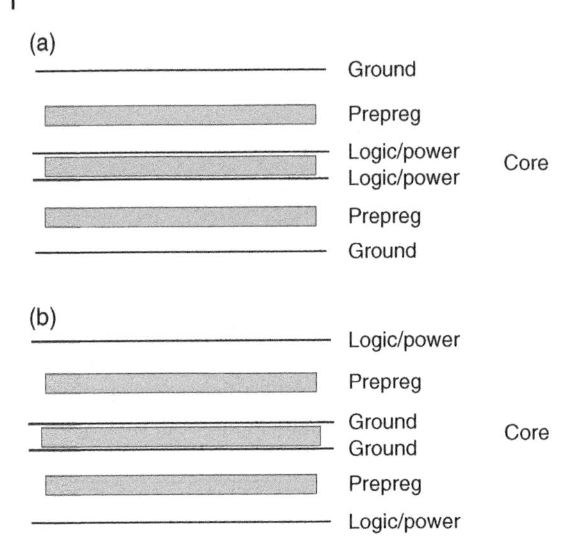

Figure 3.4 (a) and (b) Two four-layer board configurations.

There are several ways to assign functions to these layers. The core material is thick and separates the top two conducting layers from the bottom two conducting layers. This means that the outer logic layers are probably not going to cross-couple. One arrangement assigns ground to the outer layers and logic and power to the inner layers. If power is distributed on traces then power planes are not required. A second arrangement assigns ground to the core conductors and logic and power to the outer layers. This has the advantage that traces make direct connections to components. These two arrangements are shown in Figure 3.4.

Six-layer boards are built using the four-layer structure with added layers of prepreg and copper on the top and bottom surfaces. The core laminate thickness is adjusted to control the overall board thickness. The conducting layers can be used for traces, ground planes, power planes, or a mix of these functions. There are good arguments for limiting power planes to the areas under large IC structures. The power space under an IC is apt to be noisy and any logic sharing this space will couple to this interference. Power connections can always be made using traces. This geometry controls fields and limits cross-coupling. Figure 3.5 shows an acceptable layer assignment for a six-layer board.

The core layers can be ground planes and all the remaining layers can be a mix of logic and ground or power planes.

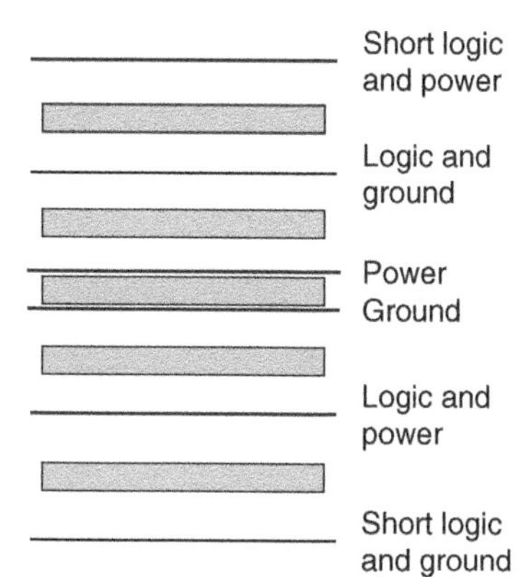

Figure 3.5 An acceptable six-layer board configuration.

3.9 Vias

Vias are plated holes that make electrical connections between conductors on different layers of a circuit board. It is common practice to connect a via to a pad at each conducting layer. Buried or embedded vias are not visible from the outside surfaces. Blind vias are visible from the outside surface but where the center holes are blocked by a conductor or plating.

The hole in a via is not a passage for fields. E field lines must terminate perpendicular to a conducting surface and can only exist if there is a potential difference. There is usually no potential difference across the walls of a small hollow conducting cylinder. It is safe to say that all wave activity and current flow involving vias will stay on their outer surfaces. Next, we treat trace transitions between layers using vias. These transitions are often a source of radiation.

3.10 Layer Crossings

When a logic line crosses between layers, the return path for current must be provided that controls the characteristic impedance. If this path is not provided, the field will spread out and the return current will use all available surfaces. The result will be reflection and radiation from the board edges. This problem is shown in Figure 3.6a. If a return path via is correctly positioned as in Figure 3.6b, this problem is significantly reduced.

(a)

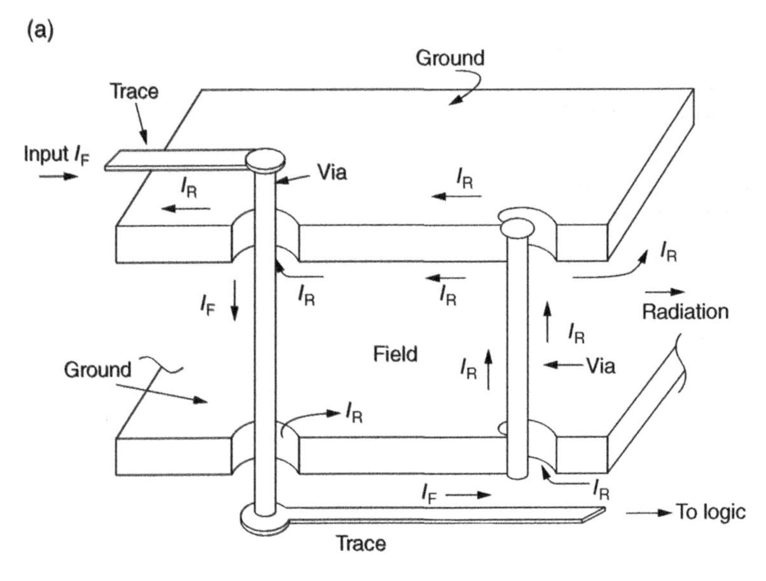

(b)

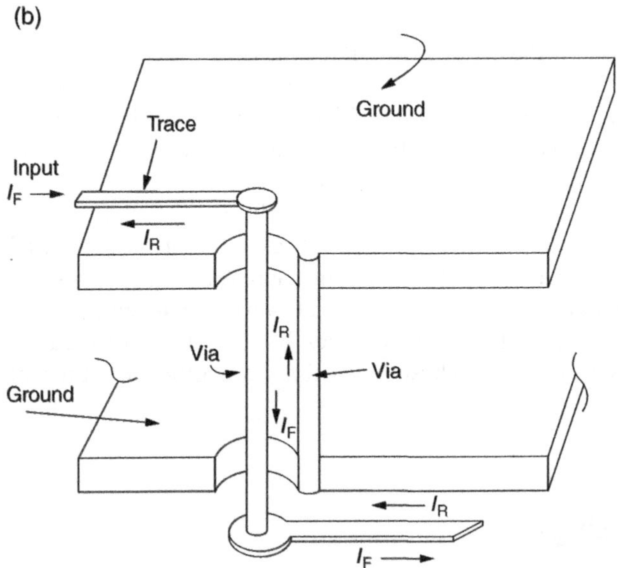

Figure 3.6 (a) A via crossing conducting planes with radiation and (b) Two vias crossing conducting planes.

The spacing between vias at a layer crossing controls the characteristic impedance of the transmission path. The characteristic impedance of parallel round conductors is

$$Z_0 = \frac{276}{\varepsilon_R^{1/2}} \log \frac{d}{r} \tag{3.1}$$

where d is the spacing between via centers and r is the radius of the via. If $d/r = 2.2$ and the relative dielectric constant is 3.5, the characteristic impedance will be 50 ohms.

INSIGHT

When a trace crosses through conducting planes, the path taken by the return current must be provided by positioning vias. Gaps must be provided at each conductor crossing. Sharing gaps invites fields to share the same space and this is direct cross-coupling.

Note that a typical trace spacing from a ground plan is 5 mils. This means that a via providing a return current path should maintain characteristic impedance to be effective. If the return current path is 1 inch away, the field will no longer be controlled and there will likely be board edge radiation. Vias are cheap and should be provided.

3.11 Vias and Stripline

Stripline is a trace between two conducting planes. Field energy is carried in parallel paths. At the transition to and from the microstrip, vias are needed to merge the two field paths into one. If the transition is handled incorrectly, some of the field energy will not merge correctly and the result can be board edge radiation. The use of vias to merge the field is shown in Figure 3.7. As rise times shorten this merging problem becomes more critical.

3.12 Stripline and the Power Plane

If stripline is placed between a ground and power plane, the merging of fields at the microstrip transition requires that the return current path use decoupling capacitors. This is a requirement at both ends of the stripline. Sharing decoupling capacitors can result in cross-coupling. The current path is usually convoluted and this can result in reflections and delays. This is the further argument for using two ground planes in stripline transmissions. This return current path in a stripline transition is shown in Figure 3.8.

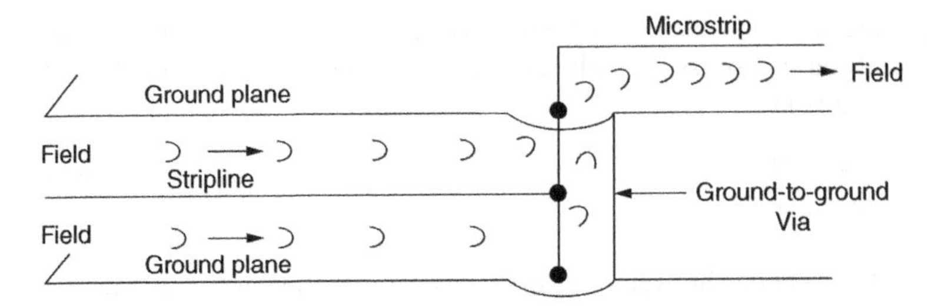

Figure 3.7 Using vias in the transition from stripline to microstrip.

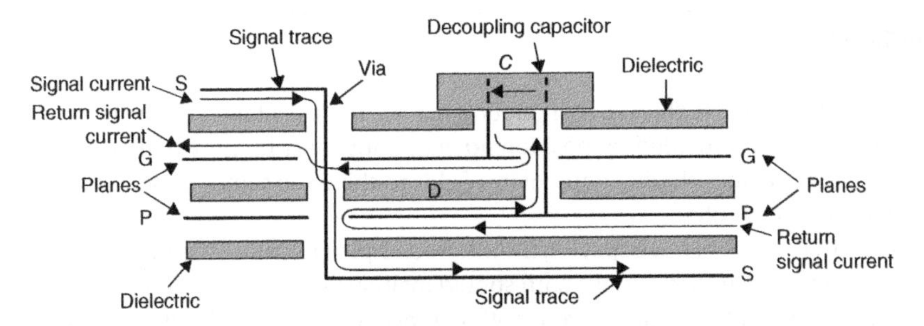

Figure 3.8 The return current path for stripline when a power plane is used.

For multilayer boards with many stripline traces, a lack of field control can result in very noisy transmissions and board edge radiation. As rise times shorten, the problem will only get worse.

3.13 Stubs

A stub is a branch transmission line usually of limited length. A stub is sometimes used to connect a logic signal to a second nearby location. If the stub length is less than two and a half times the rise-time distance there will be little impact on transmission. Beyond this distance, the reflection at the junction will reduce the forward transmission to one third and many reflections are needed before the logic level is acceptable. The preferred method is to connect the two terminal points in tandem and avoid the stub. There might be a difference in arrival time but there will be no delay caused by multiple reflections. The delay caused by a midpoint stub is shown in Figure 3.9.

The wave voltage after the matching resistor is V/2. When the forward wave reaches the stub, there are three waves: two forward waves and a reflected wave. Each forward wave has an amplitude of V/6. When a wave reaches the open end of the transmission line the voltage doubles to V/3. There are now

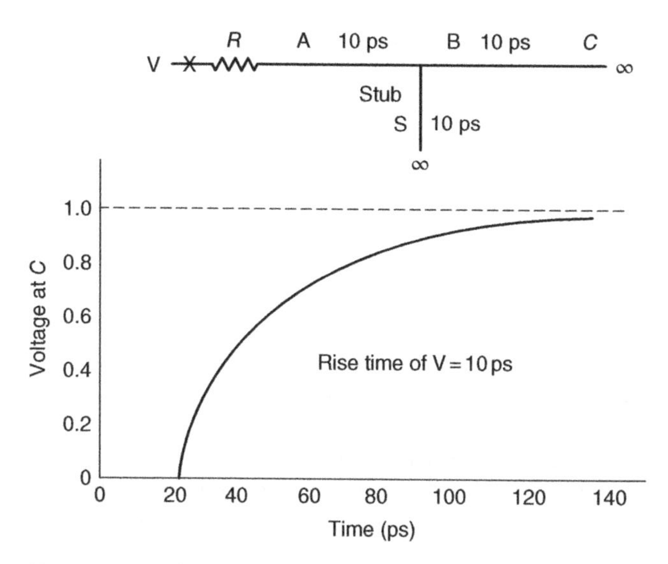

Figure 3.9 The delay caused by a stub on a transmission line.

three reflected waves returning toward the source. It will take several round trips and a dozen or more waves before the waves attenuate and a voltage of value near V reaches the two open ends. Without the stub, the arrival time is 20 ps. With the stub the arrival time is about 100 ps. If arrival time is critical then stubs are a mistake.

3.14 Traces and Ground (Power) Plane Breaks

If the return current for a transmission line must detour to get around a break in a ground plane, the break will act as a double stub and reflect energy. The only "fix" for this situation is a ground strap that completes the return current path near the trace crossing. A different layout is preferred. If the break is between a power and ground plane, a decoupling capacitor at the crossing point is needed but this situation should be avoided.

3.15 Characteristic Impedance of Traces

The control of characteristic impedance in a logic board layout is a focal point in design. In many layouts, trace lengths are short enough where terminating resistors are not required. A measure of characteristic impedance is still used to control manufacturing processes. It is an excellent quality control method that gives repeatability in both manufacturing and board performance. Controlling this one parameter does not guarantee that boards from different manufacturers will perform equally well.

There are several parameters that control characteristic impedance. For a microstrip, there are trace width and thickness, trace spacing, and dielectric constant. The manufacturer is limited in several ways. For example, trace thickness depends on available copper laminates that requires etching, plating, and cleaning. Spacings depend on the prepreg thicknesses that are available. The dielectric constant can vary with location depending on the grade of epoxy selected for cores and prepreg. This means that the board manufacturer must be consulted before parameters can be set. There are also economic considerations that must be considered. Smaller trace dimensions raise the cost of drilling holes and require tighter controls. The manufacturer must also allow for rejects in his pricing.

The following curves for different trace configurations are intended as a guide. Some of the configurations are rarely used and not all possible geometries are included. There are many computer programs available on the internet that can be used to calculate characteristic impedances. There are often variables that are not specified. Dielectrics can partially embed the trace. There are often conformal coatings that can affect the impedance. Since the dielectric constant varies with frequency, one number does not tell the full story. Step functions have a spectrum and this means that the leading edge can be affected in a nonlinear way.

The equations that are given below are approximations. Equations in the literature will vary depending on the accuracy desired and the range of the variables. The board manufacturer will have optimized his manufacturing techniques to control impedances. The curves are presented so that the reader can see how the characteristic impedance relates to various parameters. This information is not obvious by looking at the equations.

3.16 Microstrip

The traces on the outer layers of a logic board are called microstrips. The parameters for this trace geometry are shown in Figure 3.10.

The characteristic impedance of a microstrip is given by Equation 3.1

$$Z_0 = \frac{87}{(1.41 + \varepsilon_R)^{1/2}} \ln \frac{5.98h}{0.8w + 1} \tag{3.2}$$

where $0.1 < w/h < 3$ and where $\varepsilon_R < 15$.

There are several ways of presenting the relationship between parameters. Figures 3.11, 3.12, and 3.13 show the trace thicknesses of 1.5, 2, and 2.7 mils.

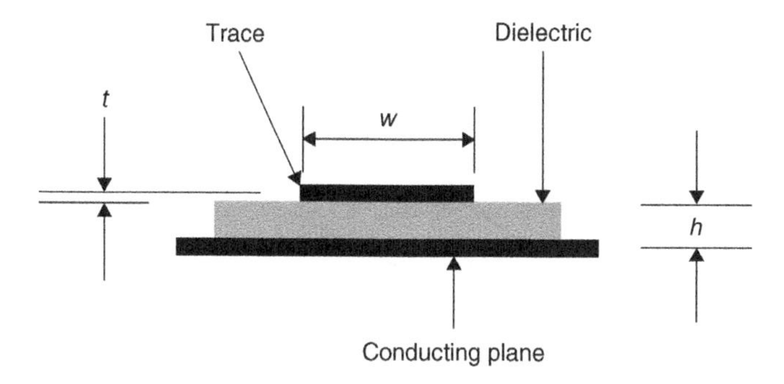

Figure 3.10 Microstrip geometry.

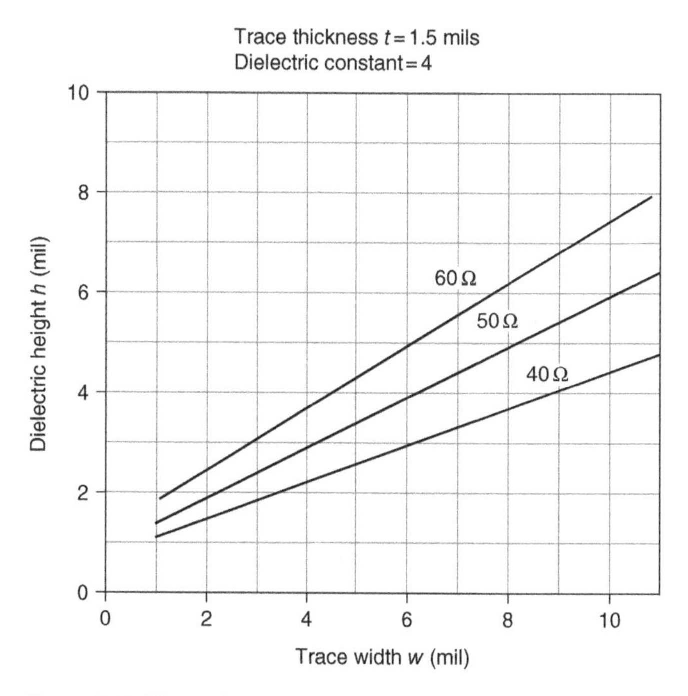

Figure 3.11 Microstrip parameters. Constant characteristic impedances for trace thickness of 1.5 mils.

In each figure, curves of characteristic impedance 40, 50, and 60 ohms are shown. The curves show that if all the dimensions are scaled, the characteristic impedance is nearly constant. The one parameter that is not controlled by the thickness of available materials is trace width.

The parameters that are used for the embedded microstrip are shown in Figure 3.14.

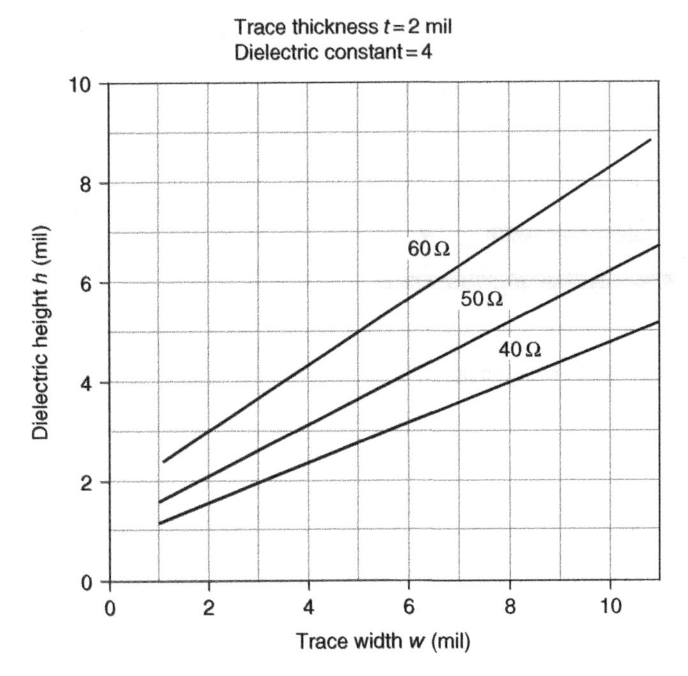

Figure 3.12 Microstrip parameters. Constant characteristic impedances for a trace thickness of 2 mils.

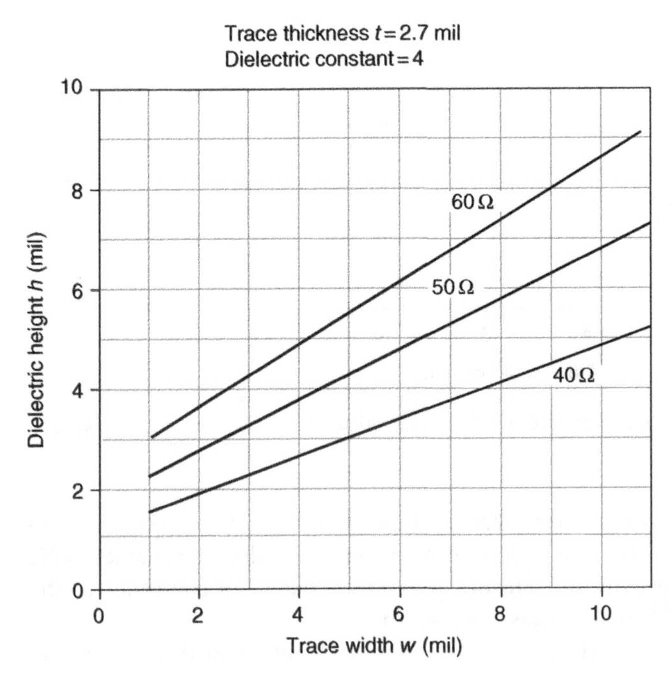

Figure 3.13 Microstrip parameters. Constant characteristic impedances for a trace thickness of 2.7 mils.

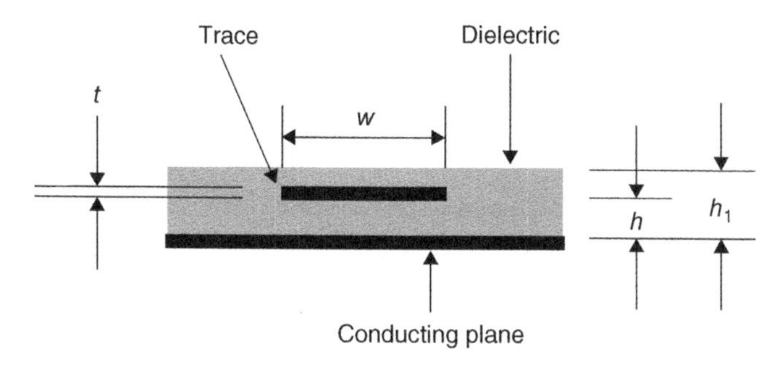

Figure 3.14 Embedded microstrip geometry.

The equations for characteristic impedance for an embedded microstrip is given by

$$Z_0 = 56(\varepsilon_R')^{-1/2} \ln \frac{5.98h}{0.8w+1} \tag{3.3}$$

where h_1 is >1.2h and $\varepsilon_R < 15$. The effective dielectric constant is given by

$$\varepsilon_R' = \varepsilon_R \left(1 - e^{-1.55 h/h_1}\right). \tag{3.4}$$

The benefit of an embedded microstrip is that surface currents for logic use the resistance of copper rather than the higher resistance of any solder plate.

3.17 Centered Stripline

Stripline refers to traces that are run between conducting planes. In centered stripline there are two paths for the energy flow. For a characteristic impedance of 50 ohms, each path has a characteristic impedance of 100 ohms.

Curves of constant characteristic impedance for a centered stripline for a trace thickness of 1.5 mils are shown in Figure 3.15.

The equation for characteristic impedance is

$$Z_0 = 60\varepsilon_R^{-1/2} \ln \frac{1.9(2h+1)}{0.8w+1} \tag{3.5}$$

where $0.1 < w/h < 2$, $t/h < 0.25$, and $\varepsilon_R < 15$.

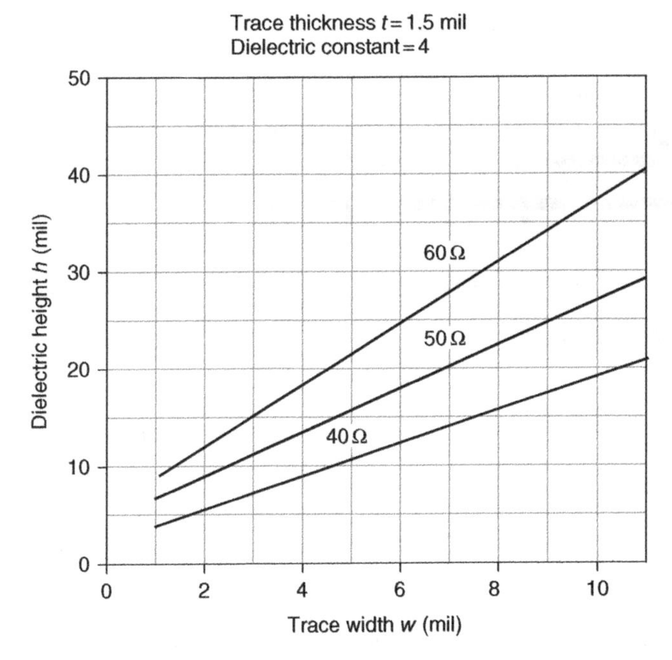

Figure 3.15 Centered stripline. Curves of constant characteristic impedance for a trace thickness of 1.5 mils.

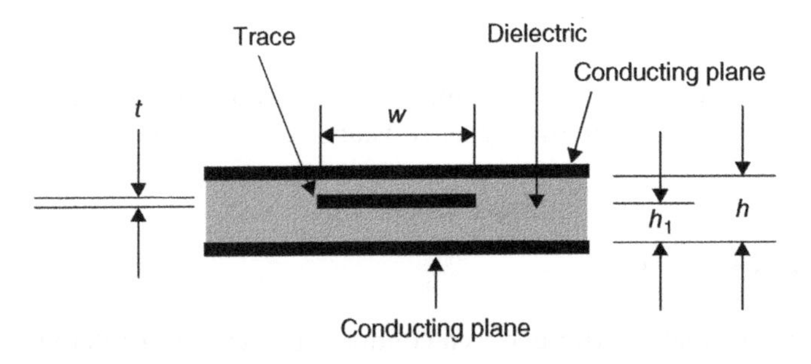

Figure 3.16 Asymmetric stripline.

3.18 Asymmetric Stripline

The parameters associated with an asymmetric stripline are shown in Figure 3.16.

The equation for characteristic impedance is

$$Z_0 = 90\varepsilon_R^{-1/2} \ln \frac{1.9(2h+t)(4h_1-h)}{4h_1(0.8w+1)} \tag{3.6}$$

where $h_1 > h$, $0.1 < w/h < 2$, and $t/h < 2.5$.

3.19 Two-Layer Boards

It is practical to use a two-sided board for fast logic. Ground and power planes are replaced by a grid of traces that carry both power and signal. Every logic line consists of three parallel traces called a triple. The triple consists of a logic trace between a ground and power trace in either order. The top of the board carries the triple in the x direction and the bottom of the board carries the triple in the y direction. Vias are used to transition the triple from the x to the y direction. On a 1/16-inch board for a trace width of 10 mils, a trace separation of 10 mils, and a trace thickness of 2.7 mils, the characteristic impedance is about 50 ohms. Decoupling capacitors are placed across the triples at each IC. Care must be taken to add enough copper to feed power to the grid of triples. The grid supplies power and logic to each IC. A ring should connect all the grid ground and power traces together at each IC. Any board areas not used can be flooded with ground (see Figure 3.17).

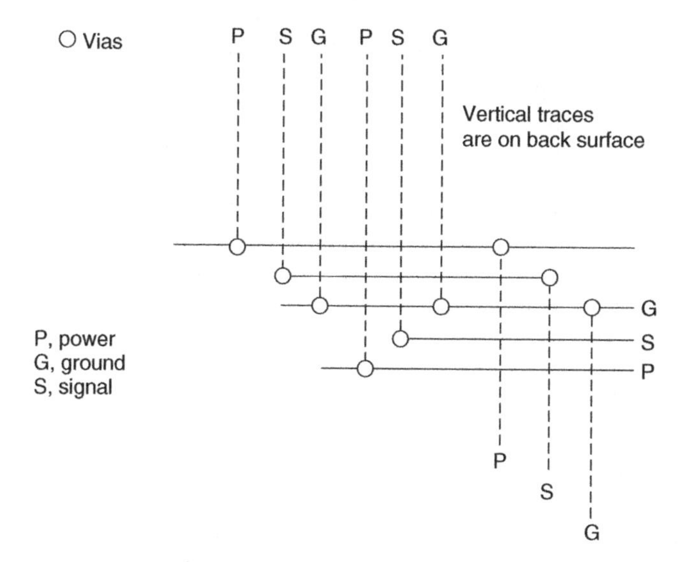

Figure 3.17 Trace pattern for use on a two-sided board.

3.20 Sine Waves on Transmission Lines

There are many applications where sine waves are processed on a circuit board. An application might involve a communications link. The role of characteristic impedance is somewhat different in these applications. In logic it was important to transfer a logic state in a short period of time. In a communications link, it is important to deliver a steady flow of sinusoidal energy to an antenna.

What happens to the first cycles that reach the antenna can be ignored. The steady state energy that reaches the antenna is important.

Step waves gave us a unique understanding of how energy is managed on transmission lines. The effective handling of logic requires that reflections, if they occur, are handled correctly. In logic, one round trip should complete wave activity. An ideal transmission line has no losses. As we have seen, there can be step-wave oscillations on a transmission line that is open or short circuited. Without resistance there is no way to lose the energy and it reflects back and forth. This means that an unterminated transmission line connected to a voltage generator cannot draw any real energy after the line is once energized. This is also true if the line is terminated in a capacitor, inductor, or any reactive network.

For a sine waves generator and an unterminated transmission line, there may be reactive energy flow where equal amounts of energy are added and subtracted on each cycle. This is exactly what happens when a sine wave voltage generator is placed across a capacitor or an inductor. A shorted or open transmission line looks like a capacitor or inductor that depends on line length and frequency. There are reflections at both ends of a line that obey the rules that were developed in the first chapters.

When a transmission line is used to carry sine wave energy to an antenna, reflections are not desired. The hope is that the arriving energy enters the antenna and radiates. Antennas must couple to the impedance of free space or 377 ohms and transmission lines on boards are typically 50 ohms. The problem of matching impedances and optimizing radiation is treated in Chapter 5. It is worth commenting that the impedance of free space is a resistance. Energy cannot be radiated into a reactance.

3.21 Shielded Cables

Shielded cables when used in microphones are a braid made from woven tinned small copper wires. This cable often carries conductors that supply power for preamplification. Wireless technology has largely replaced the standing microphone and the need for a cable. This technology has also replaced the cable connectors located at the receiving hardware. Power for the microphones is supplied from small batteries.

Instrumentation for strain gauges requires up to 10 conductors. Excitation and remote sensing takes up 4 lines, signal and calibration lines take up 5 lines, and the guard shield makes 10 conductors. The shielding for this cable is often aluminum foil that is anodized on the inside surface. The foil is folded along the cable and does not allow surface currents to circulate radially. A braided drain wire is run along the outside surface of the cable making it practical to terminate the foil at the cable ends. The shielding against interference fields below 100 kHz is excellent. For analog work, any rf coupled to the inner conductors can be filtered passively at the instrumentation.

Cables that connect piezoelectric transducers to signal conditioning must treat the triboelectric effect. In applications with a high ambient noise level, rubbing of the braid against the dielectric generates unwanted signals. Cables that limit this noise are available from transducer manufacturers.

3.22 Coax

The need to control the characteristic impedance in the transport of fields requires controlling the conductor geometry. Shielded cables that control the characteristic impedance are called coax. The geometry can be controlled by using a dielectric to center the inner conductor. The characteristic impedance of a coaxial transmission line with an air dielectric is shown in Figure 3.18.

When a coaxial cable is needed for a transmission over any distance, the use of a dielectric allows a distributed reflection that is undesirable. To solve this problem, the inner conductor is centered by a spiraling nylon cord. The surface of braid also causes a distributed reflection. The preferred outer conductor is a thin smooth-wall conductor that can flex. The copper center conductor is an alloy that will not easily bend or kink. A corrugated outer conductor can add flexibility but it adds to the distributed reflection and limits performance at high frequencies.

3.23 Transfer Impedance

Braided coax has a limitation that is caused by the conducting braid. Current on the inner surface will tend to follow individual strands of copper to the outer surface. In a like manner interfering external fields cause current flow to follow the individual strands into the cable. Any current on the inside surface is field and this is interference. This interference travels in both directions and appears as a voltage on the center conductor. This coupling can be reduced by using a double braid, but the added attenuation is not that useful.

The ratio between half the open-end voltage and the shield current flow is called the transfer impedance of the cable. Figure 3.19 shows a typical terminated cable under test.

The transfer impedance for several cable types is shown in Figure 3.20. Note that thin wall at high frequencies provides excellent shielding where braid fails above 100 MHz. A transfer impedance of −40 dB ohms means that the transfer impedance is 0.01 ohms/m. A current of 1 A will develop an output voltage of 50 mV in 10 m of cable.

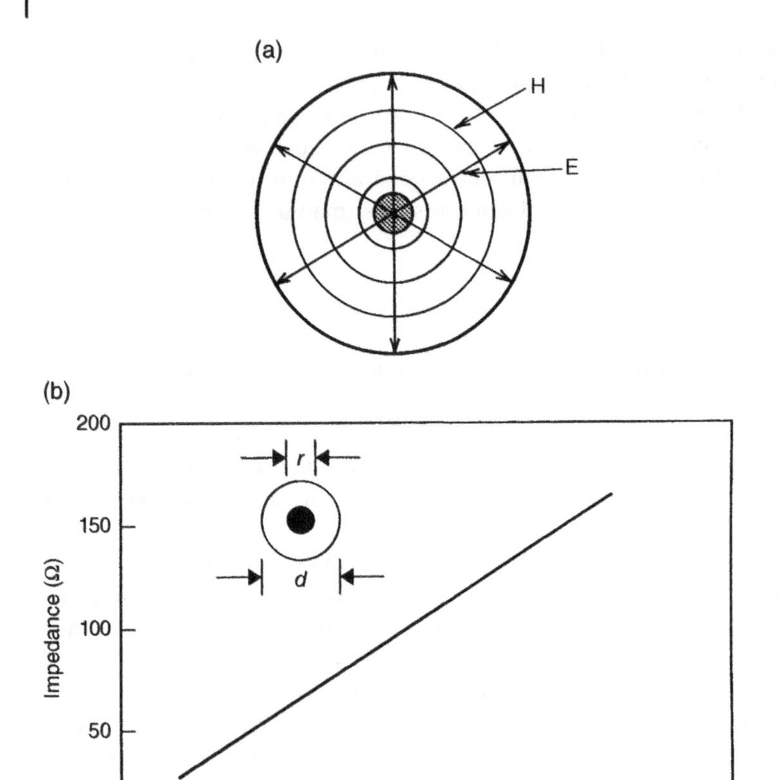

Figure 3.18 The characteristic impedance of a coaxial geometry.

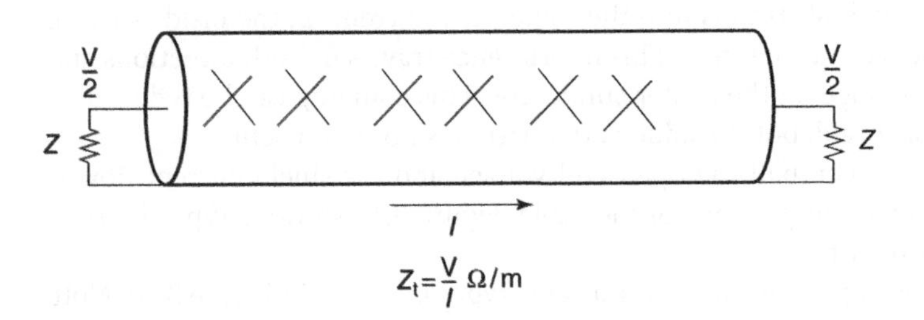

Figure 3.19 Transfer impedance test for a coaxial cable.

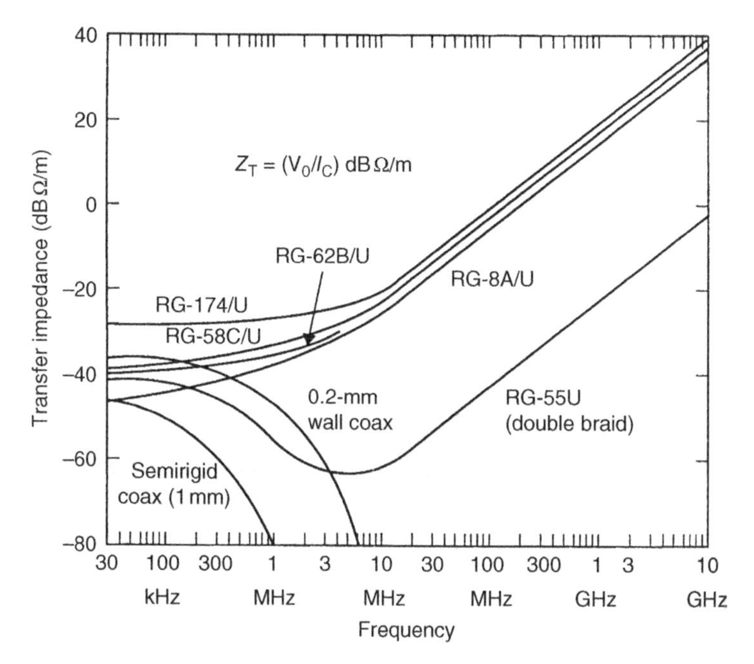

Figure 3.20 The transfer impedance for several standard cables.

The important thing to note is that the shielding of braided cable fails to be effective above about 300 MHz. For distances under 30 cm and frequencies under 1 GHz, the type of cable shield is usually unimportant.

The ideal termination of a shielded cable provides a smooth transition of interference current from the cable shield to the connector shell to the mounting surface. When the interference levels are severe, the best practice is to use backshell connectors. There must be a 360° shield termination without gaps. The connector body must be bonded electrically to the mounting enclosure preferably with a conducting gasket. For many applications where circuit boards are not enclosed, cable shields are used to control characteristic impedance not to limit interference. Closure does not necessarily control characteristic impedance.

Cables are often used to interconnect power supply leads, control leads, indicators, as well as logic. If any of these leads carry any interference current, then there is a good chance of cross-coupling and the logic will be compromised. Here is a story that will illustrate this point.

I received a call from a friend on a Saturday morning. He had a piece of new hardware that had to be shipped. He had worked late the previous night, but had made no progress in solving an interference problem. This hardware measured oil flow and every time the pumps were operated there were errors. I got in my car and drove the 30 miles to see if I could help. He had connected a separate control panel to a computer using a shielded cable. I looked at the signal list and I found that he had added a conductor that connected the equipment ground of the control panel to the equipment ground of the computer. We cut this connection and the problem was solved. Normally transient effects would cause currents to flow on the outside cable shield. This added conductor brought the interfering field into the cable.

3.24 Waveguides

Transmitting high power at high frequency requires both a low characteristic impedance and a high voltage. In Figure 3.18, a low characteristic impedance in coax requires the spacing between conductors to be small. At a spacing of 5 mils, a 5-V signal already has a voltage gradient of 1000 V/inch. This means that coax is not going to be effective for high-power transmissions.

A waveguide is a hollow conducting cylinder like a rectangular air duct. At a frequency where one dimension of the opening is a half wavelength, sine wave electromagnetic energy will propagate along the cylinder without a center conductor. At specific higher frequencies, a waveguide will support propagation. A good example of this wave action is the operation of a microwave oven. A klystron supplies energy to a waveguide terminated in the oven compartment. The load in this case is the food in the oven. Without a center conductor, the field voltages in a waveguide can exceed 100,000 V/m. Waveguides are used in radar and in supplying energy to UHF transmitters.

As the frequency is raised the number of modes that can be carried on a microwave link increases until there is a continuum. The lowest frequency that can be transported is called the cutoff frequency. Driving the waveguide below this frequency is said to be driving the waveguide beyond cutoff.

The attenuation of wave energy beyond the cutoff frequency is given approximately by

$$\mathrm{A}_{\mathrm{WG}} = \frac{30h}{d} \tag{3.7}$$

where d is the largest aperture dimension, h is the aperture depth, and A_{WG} is the wave attenuation in decibels.

A seam that is a few millimeters wide and 30 cm long will not be effective in shielding high-frequency energy as A_{WG} is unity. This is why a gasket is needed to seal the door on a microwave oven. The gasket must make ohmic contact with the entire seam so that surface currents can flow freely across the boundary.

FM radio operates on a carrier at about 100 MHz. The half wavelength is about 1.5 m. FM radio signals easily propagate into a road tunnel or an underground parking garage. An added conductor is needed to bring an AM signal into these same structures.

3.25 Balanced Lines

A balanced transmission line is a pair of conductors surrounded by a shield. Each conductor and the shield is a 50-ohm line. The shield is a shared return path for two transmission lines. The term balanced means that the two transmission paths have the same characteristic impedance.

INSIGHT

For a balanced line, the field pattern for each transmission path is unsymmetric. This means that the return current for each transmission path tends to concentrate on the shield surface where it is closest to the forward current. It is important to realize that fields do not combine their energy even though they share the same space. If a termination is required, two resistors are needed for the two transmission paths.

3.26 Circuit Board Materials

Board manufacturers have many choices in the glass epoxy used for board manufacture. Many forces have been at work to change board material. Increasing clock rates have brought on the need for a finer glass weave. Traces that by chance were located over a single glass ribbon would have a different characteristic impedance. One way to average the character of the weave is to use two layers of prepreg. This raises the cost but usually stabilizes the characteristic impedance. One of the biggest changes in board manufacturing occurred when lead was removed from solder. This directive is known as RoHS which stands for the Restriction of Hazardous Materials. Circuit boards must be cured at a higher temperature and this requires changes to the epoxy resin. Another problem is that smaller trace size requires more accurate drilling. Drilling and the use of thinner layers has increased the issue of mechanical stability. Many products require that boards be halogen free. If there is a fire, the gases that result could be lethal. Prepreg must match the mechanical properties of the copper laminates. All of this variability has made the problem of inventory difficult. It also leaves very little opportunity for a company to experiment with different materials.

Board manufacturers build products for many different users. Some users worry about dielectric losses while others worry about signal delays caused by variations in the dielectric constant. In most logic boards these are not critical parameters.

A board designer must be aware of these issues and should consult with board manufacturers so that a manufacturing specification can be agreed upon. As mentioned earlier, boards from different sources my not function the same way.

Problem Set

1. Describe the wave action when a voltage is switched onto a 100-ohm transmission line in series with a 50-ohm line.
2. A capacitor of 1 nF has a natural frequency of 100 MHz. What might the characteristic impedance be?
3. A 50-ohm line connects to a 100-ohm resistor. What is the reflection coefficient?
4. A 50-ohm transmission line is charged at 10 V. A wave of 1 V is added to the line. What is energy in this wave? Assume the line length is 30 cm.

Glossary

Balanced line Two conductors carrying odd mode logic. When one conductor is at logic 1, the other is at logic 0. The average value is kept at ½ (Section 3.25).

Coax A shielded single conductor cable with a controlled geometry intended for high-frequency applications (Section 3.22).

Cross–coupling The unwanted transfer of energy between two controlled energy paths.

Cutoff frequency The lowest frequency that can propagate on a waveguide (Section 3.24).

Decoupling capacitor A capacitor in a logic circuit intended to supply local energy to transmission lines (Section 3.2).

Die The core of a semiconductor product. A group of interconnected logic switches and semiconductor components usually made from silicon that performs a set of logic functions.

Distributed reflection The reflection of wave energy along a cable. This reflection can result from irregular conducting surfaces or from variation in the dielectric constant (Section 3.22).

Embedded energy The energy between conducting planes on a circuit board. The energy on transmission lines with a conducting plane as the return path (Section 3.3).

Energy The state of a system that allows it to do work. Energy is stored in electric and magnetic fields. It is stored in a mass based on its position in a gravitational field. It is stored in a mass based on its velocity. A 5 V signal in transit on a 50-ohm transmission line consumes ½ W of power. Energy is power times time. In 0.1 μs the energy moved is 0.5×10^{-7} joules (Section 3.2).

ESD Electrostatic discharge.

Glass epoxy The board material in common use in the manufacture of circuit boards. The term "glass" refers to a woven glass fabric used in the manufacture. The fineness of the glass mesh determines how constant the dielectric constant will be.

Ground A conducting surface used as a local zero of potential. It can also be earth, power neutral, or a wire. The term has a very specific meaning in utility power.

Ground plane A conducting surface on a circuit board at zero potential. The return current path for transmission lines (Section 3.3).

IC Integrated circuit.

Microstrip Traces on the outer surfaces of a circuit board (Section 3.16).

Power The unit of power is watt. The rate at which energy is moved. A joule per second is 1 W. A 5-V signal on a 50-ohm line carries ½ W (Section 3.2).

Power plane A conducting surface on a circuit board at the power supply potential (Section 3.3).

Prepreg Sheets of partially cured glass epoxy used in board manufacture. Sheets of prepreg comes in many standard thicknesses (Section 3.8).

Radiation The transfer of electromagnetic energy into free space (Section 3.7).

Reactance The opposition to sinusoidal current flow in capacitance or inductance (Section 3.4).

Return path In a transmission line, the path taken by current to return to the driving point. Vias provide both signal and return current paths.

Shielding The use of conductors to contain an electric or magnetic field. Shielding can be a braid on a cable, a conductive housing, or a conductive layer in a transformer. It can also be a layer of magnetic material (Section 3.21).

Stripline Traces located between conducting planes on a circuit board used as transmission lines. Centered lines are preferred as the energy is divided equally in two paths (Section 3.11).

Stub A short transmission line usually connected to a main transmission line at some midpoint. A stub can be used to terminate a transmission line. Stubs can be terminated or used open or short circuited (Section 3.13).

Transfer impedance The ratio of voltage coupled into a shielded cable from a current flowing on the shield's outer surface. Because the coupled wave reflects and doubles at the open ends, the measured voltage is divided by 2 (Section 3.23).

Vias Plated through holes on a circuit board that connect traces located on different conducting layers. Buried vias are not visible from the outer surface.

Blind vias have the drilled holes filled with solder. Vias are supported by pads at each layer. Fields do not penetrate the holes associated with vias (Section 3.9).

Waveguide Any hollow conducting cylinder that propagates electromagnetic energy without a center conductor. Usually used for sine wave propagation at high-power levels at frequencies above 500 MHz (Section 3.24).

Waveguide beyond cutoff A waveguide driven at a frequency that is too low to propagate energy.

Answers to Problems

1. The voltage V sets a wave in motion that reaches the 50-ohm transition. The transmission coefficient is 0.667 V. The reflection coefficient is −0.333 V. When this wave reaches the voltage source the reflection reverses the polarity and a wave of +0.333 V moves forward. When this wave reaches the transition, two new waves are generated. The forward voltage thus increases in steps that reach V in the limit.
2. Assuming the entire capacitance resonates with the entire inductance, then $f_n = 1/6.28(LC)^{½}$. The inductance is 2.54 nH. Using this inductance the characteristic impedance is $(L/C)^{½} = 1.2$ ohms.
3. 0.333.
4. A wave takes 1 ns to travel 30 cm. At 10 V and 50 ohms the power level is 2 W. The energy in a ns is 2×10^{-9} joules.

4

Interference

4.1 Introduction

Radiated interference can be roughly divided into two parts: energy that couples within a circuit and energy that enters from outside sources. Entering interference can come from radiation or cable connections. We discuss lightning and electrostatic discharge (ESD) separately.

This chapter deals with interference of all types and ways to limit its impact. Errors can occur in the sampling process. Utility power can be a source of interference when there are voltage surges, voltage dropouts, neutral voltage drops, and capacitive coupling. There can be instabilities in active circuits. Logic errors can occur when there are reflections, insufficient signal, or a lack of energy to carry the logic.

High-speed logic requires greater bandwidth and this adds to the radiated levels. Reducing operating voltages limits radiation but makes circuits more susceptible to interference. The use of hard wires to carry signals through any distance has always been a problem. One difficulty is the coupling to interfering fields. The use of fiber optics and high-frequency carrier links has eliminated many problems but this comes at a price. The data must first be put into a form where it can be transmitted.

We have already discussed two sources of interference generated on a circuit board: radiation from traces at the leading edges of waves and board edge radiation caused if fields are not provided a controlled space to use. The cross-coupling of logic signals is another type of interference that is controlled by the layout. Cables that interconnect devices can couple to both local and remotely generated fields. This is nature taking every opportunity to spread energy out over any available conductor geometry.

When energy leaves the confines of transmission lines, the waveforms lose their step function character. The best way to measure and analyze this interfering is to use a sine wave analysis. The rise time character as discussed in Appendix B, shows that the frequency of interest for an analysis is given by

Fast Circuit Boards: Energy Management, First Edition. Ralph Morrison.

$1/\pi\tau_R$ where τ_R is the rise time or transition time. The approach that is taken is to assume that the interference can be represented by a sine wave signal at this rise-time frequency and amplitude. The amplitude of the response using this one frequency represents the degree of coupling or interference. This approach can be used to estimate the effect impulses such as ESD or lightning can have on a circuit. This same approach can be used to estimate the impact logic signals will have in cross-coupling.

As an example, an ESD event can be characterized as a 5-A pulse. A typical pulse discharges field energy stored within a distance of 10 cm in about a nanosecond. The rise-time frequency is about 300 MHz. A lightning pulse can be characterized as a 100,000-A pulse. The field energy that is discharged and radiated is stored within a few hundred meters of the ionization path. The time for the initial energy to cross this distance is about 2 μs. The rise-time frequency for this event is about 650 kHz. In both cases, the current level can be used to calculate an H field at a distance *r*. The B field in space is simply μ_0H. The induced voltage in any nearby loop depends on the changing B field, the loop area, and the frequency (see Equations 1.16, 1.18, and 1.19).

N.B.

Fields can damage circuits. An interference current does not need to flow in the circuit to do damage. The coupling of field energy is proportional to loop area. Field intensity falls off proportional to distance, not distance squared.

4.2 Radiation—General Comments

Radiating structures (antennas) can be dipoles, loops, horns, slots, parabolic dishes, or arrays. Transmission lines and waveguides are used to carry energy from a circuit to these antennas for radiation. These same antennas can couple to fields that bring information into a circuit for processing. At 1 MHz, a transmitting antenna that is a quarter wavelength long is 75 m tall. At 300 MHz, a quarter wavelength in space is 25 cm. For practical reasons, receiving antennas are often much shorter than a quarter wavelength. The coupled voltage is roughly proportional to antenna length being optimum at a half wavelength.

A trace over a conducting plane forms a rectangular loop antenna that has a length equal to the distance traveled in one rise time. The field present along the rest of the trace is unchanging and it is therefore not radiating. The E field radiation at a distance of 1 m from a square centimeter of radiating area for a 1 V-rms signal on a 50 line at 100 MHz is about 66 dBμV/m. This one example of radiation can be used to estimate the radiation from an entire circuit board. Radiating efficiency increases with the square of frequency. Field strength falls off linearly with distance. On a circuit board, the radiation that can be controlled comes from traces. Here the loop areas are small but the number of active loops can be in the hundreds. Radiation from board edges is usually the result of errors in the layout (see Sections 3.10 and 3.11).

Radiation levels from sources having the same spectrum are additive. Radiation levels from sources with a different spectrum are summed by taking the square root of their squared values. As an example, 30 mV of radiated carrier and 40 mV from a logic source will sum as $(30^2 + 40^2)^{1/2} = 50$ mV of interference. The sum can then be given in terms of decibels once a reference level is selected (see Appendix B).

4.3 The Impedance of Space

In Chapter 2, Equations 2.5 and 2.6 related the permeability and permittivity of free space to the ratio of current and voltage. It is easy to show that

$$\left(\frac{\mu_0}{\varepsilon_0}\right)^{\frac{1}{2}} = Z_0 \tag{4.1}$$

where Z_0 is called the characteristic impedance of free space. Using the values given in Section 2.5, this impedance is 377 ohms. This measure is also the ratio of the E and H field in an electromagnetic wave at a distance from a radiating source. Near the radiator, the wave impedance depends on the conductor geometry. For sine wave radiation, the distance λ traveled in one cycle is called a wavelength. The distance $\lambda/2\pi$ is called the near-field/far-field interface distance. Beyond this distance, radiated waves are called far-field waves. At 1 GHz, a wavelength is 30 cm and the interface distance is 4.7 cm. For a logic square-wave signal, the interface distance is different for each harmonic. In the near field, coupling to nearby circuits will vary depending on the wave impedance of each harmonic. In the far field there is equal energy in the electric and magnetic fields.

4.4 Field Coupling to Open Parallel Conductors (Sine Waves)

Field coupling to a simple conductor geometry (wires) is shown in Figure 4.1. There are two things to note. Field coupling is optimum when the interfering field moves parallel to the path take by the conductors. The coupling is proportional to both path length and conductor spacing. Coupling is therefore proportional to loop area. The maximum distance that can be used in a calculation is the distance a wave travels in one clock time. For a larger dimension the

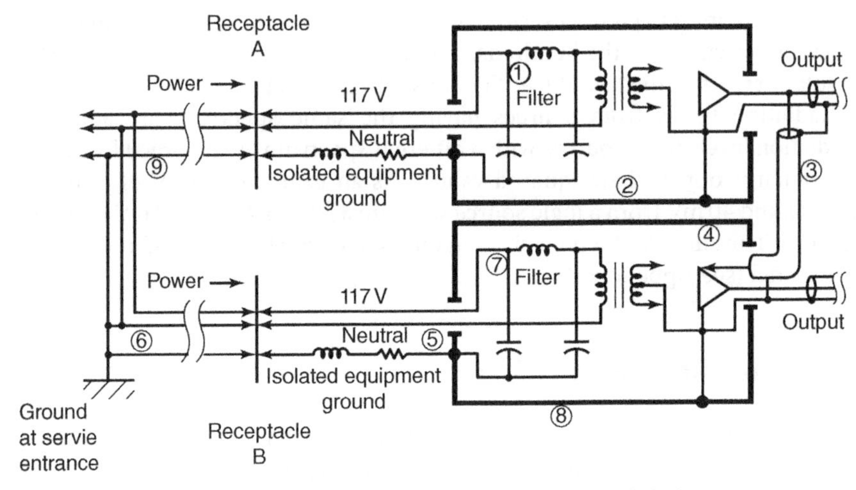

Figure 4.1 Field coupling to parallel conductors (wires).

coupling lessens but it is not a good idea to rely on any cancellation. The simplest way to reduce coupling is to reduce the coupling loop area. The interference fields can be from a nearby circuit board or from a remote source.

In Figure 4.1, there are two field coupling areas. The first is the area between the two top conductors and the ground plane. In logic, voltage coupled into this area is called even-mode interference. In circuit terminology, voltages that are coupled into this loop area are called common-mode interference.

There can be field coupling to the area between a pair of conductors carrying a signal. In logic, this coupling is called odd-mode interference. In circuit terminology, this is called normal-mode interference. Depending on the cable type, this coupling can be reduced by twisting the conductors in the cable or by shielding.

4.5 Cross-Coupling

There are designs that require long transmission paths where energy can cross-couple causing interference. This section reviews the basic process so that a designer knows what to expect. There are many factors that should be considered including line lengths, trace proximity, logic direction, logic polarity, line terminations, rise time, and the logic family. Cross-coupling is really the transfer of energy but the only coupling mechanisms we have discussed involve the E and H fields treated separately. We discuss coupling in terms of voltage as this is what we usually measure.

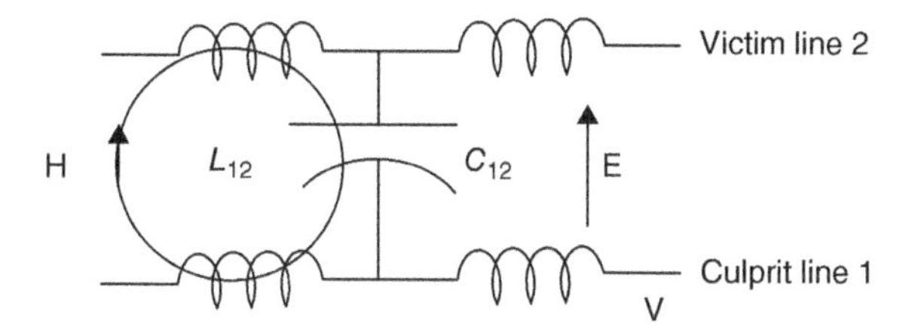

Figure 4.2 The mutual capacitance and mutual inductance between two transmission lines.

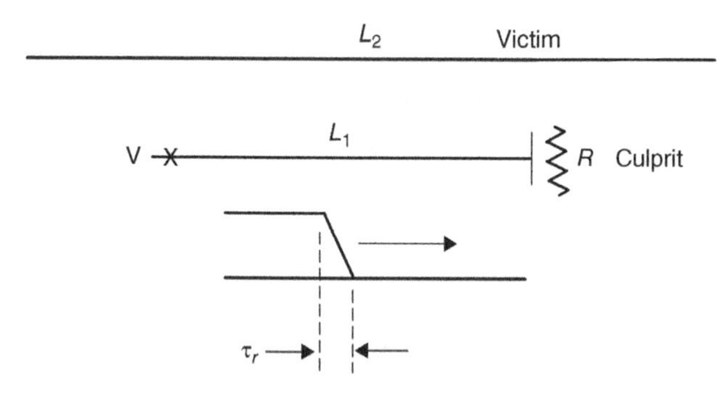

Figure 4.3 A step-function wave applied to a culprit line.

The circuit we consider is shown in Figure 4.2. The figure represents two transmission lines with mutual inductance L_{12} and mutual capacitance C_{12} per unit length.

The two transmission lines are L_1 (the culprit line) and L_2 (the victim line). These lines are shown in Figure 4.3.

As the culprit wave progresses to the right, the only area where the magnetic field couples to the victim is during the rise time. The coupled energy is restricted as the forward wave cannot exceed the speed of light. The coupled wave that moves to the left reaches a maximum amplitude in the distance traveled during the rise time. The rise time of the coupled energy is assumed to be twice that of the culprit rise time.

INSIGHT

The problem in describing this wave action is that the culprit wave and the coupled wave cannot have a relative velocity of $2c$ as both waves carry energy. Any observer will see a velocity of c or less. The important observer is the receiving logic.

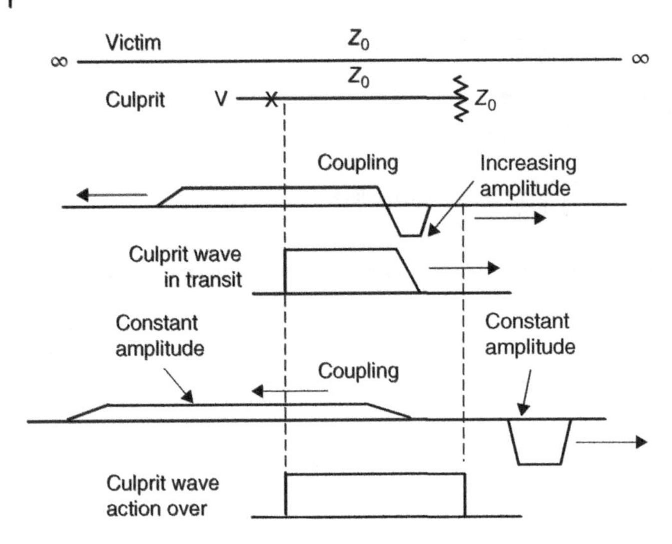

Figure 4.4 Inductive coupling between transmission lines.

The wave component that results from inductive coupling is given approximately by

$$A_R = V\frac{L_M}{4L} \tag{4.2}$$

where L is the inductance per unit length in the two transmission lines and L_M is the mutual inductance. The wave action described above is shown in Figure 4.4.

The voltage component that results from capacitive coupling is similar to the inductive component. The total reverse-coupled crosstalk wave amplitude is given by Equation 4.3, where C is the self capacitance and C_M is the mutual capacitance per unit length.

$$A_R = V\left[\frac{L_M}{4L} + \frac{C_M}{4C}\right]. \tag{4.3}$$

4.6 Shielding—General Comments

In Chapter 1, the fundamental idea of shielding was introduced. Potential differences between conductors imply electric fields. If fields are fully contained by a conductor geometry, then the circuit generating this field is shielded. This basic idea leads to shielding methods that include metal enclosures, shielded cables, gaskets, and passive filters. The intent is to control which fields can leave or enter a controlled space. Power line filters on hardware allow utility power to enter but block high-frequency energy. In general, line filters must be electrically outside but physically inside a metal enclosure. This requires the proper partitioning of space. Energy uses conductor pairs as transmission lines and nature pays no attention to labels. The NEC prohibits placing any component in the equipment grounding path, so this lead must remain electrically "outside" of any enclosure.

In the real world, circuits do not function in isolation. Even a cell phone must receive and radiate energy. Many circuits receive signals and operating energy using unfiltered conductors. All unfiltered conductors that enter a controlled space will bring in external field energy and provide a path for internal fields to exit the space. Field energy can enter and leave a controlled space through apertures and seams. The game that must be played is to limit the flow of interference energy and provide a clear path for desired energy. In analog work, the approach that is taken is to shield low-level signals until they can be amplified and the impedance can be reduced to a few ohms. Active circuits can be used to reject any interference that is common-mode. Out of band interference can be removed by passive filtering.

4.7 Even-Mode Rejection

In logic, a signal often must be sent between boards or different pieces of hardware. It is expected that there will be a potential difference between circuit commons even if they are ohmically connected. The field in the intervening space will couple to all conductors that cross through this space. To limit this interference, a balanced signal can be generated in the transmitting logic. Both signals will couple to the same interference. The receiving circuit can amplify the difference in potential which eliminates the ground potential difference. This ground difference of potential is called even-mode interference. The arriving signal is usually biased at the receiver so that it is unipolar. The ground difference of potential is called even-mode interference.

In applications where balanced logic is used, the transmission distance requires terminations to avoid voltage doubling. The cable is usually a balanced 100-ohm pair that ends on 50-ohm lines on the receiving board. No terminating resistors are used at this point. The two signal lines are separated on the receiving board so that the two fields use a different space. At the receiving logic, two 50-ohm terminating resistors are required. This circuitry is shown in Figure 4.5.

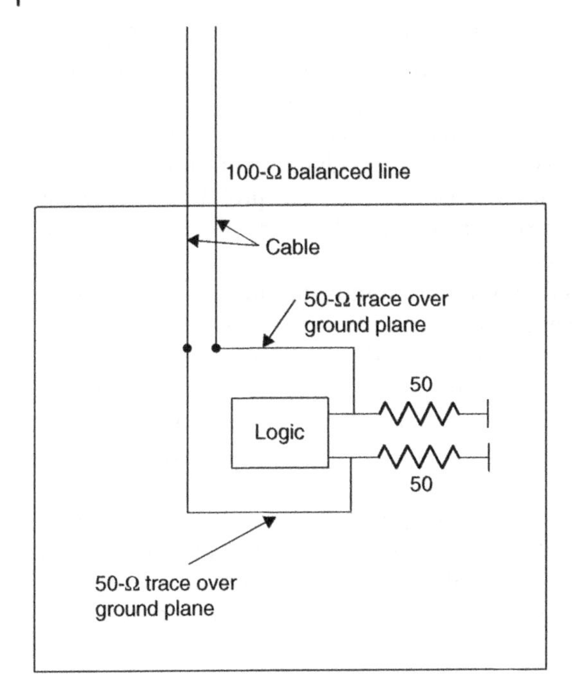

Figure 4.5 The termination of a balanced transmission line.

4.8 Ground—A General Discussion

In circuit board designs, the common or reference conductor at 0 V is often called ground. In many cases there is no earth connection. Obvious cases are cell phones and circuits in automobiles or aircraft. A ground can be as small as a postage stamp or as vast as an ocean.

There is often comfort in a concept that has no real basis: the idea that there is a "ground" that is *quiet* and that if we can connect to it, all noise problems will be conquered. The feeling is that the earth is a sump for noise. It can "absorb" the noise so that it will not reappear in a circuit. The circuit theory idea that current must flow in a loop is ignored. It is easy to find the entry point but the exit point(s) are a mystery. This feeling is so strong that significant attempts are often made to "ground" circuits to earth using tons of copper rods. The term *quiet* implies that the buildings will be noise free and will not allow radiation. Of course there is no physics to support this contention. Earth connections are important in power safety and in lightning protection. There are codes for the transport and grounding of utility power. The fact that these same conductors can function as transmission lines for rf energy is not considered. My trite remark is that nature does not read labels or color codes.

A bit of history will help explain a few common misconceptions. In the early days of electrification there were many cases where lightning entered a facility on power wiring. There were fires that resulted from a lack of fault protection and there were shock hazards. Under pressure from banking, insurance companies, and help from the National Fire Protection Agency, the National Electrical Code (NEC) was written and made a requirement in all electrical installations. This code is not a government regulation although it is adopted as a rule by most municipalities. The code defines the meaning of key words. *I have, in a few cases, indicated these words in italics.* The code is amended every 3 years to accommodate new practices and new materials. The latest code is dated 2017. The code provides rules and makes no attempt at any analysis. The code requires that utility power be connected to earth at the *service entrance* at each facility. The assumption is that lightning that strikes utility conductors will flow to earth outside of a facility. This rule applies to underground power entrances. It is important that all facilities should obey the same rules. Not all countries follow these same grounding rules. On board naval ships, power is *ungrounded.* In some areas on a ship, a distribution transformer is added and the secondary is *grounded.* Floating the power system reduces electrolysis problems and allows for faults in combat that could disable key equipment. Unfortunately when power is floating, any changes in power demand produces interference that propagates over the entire power system even if it is in conduit.

Many facilities on the same utility line make *neutral* connections to earth. In three-phase distribution, the phase loads are not exactly balanced. This means that some *neutral* current uses the earth as a return conductor. Some of this current in turn uses gas, water, and telephone lines as well as building steel. As pointed out, this current only flows because there are fields. The utility has no legal obligation to reduce this current. Generating this current requires capital equipment and balancing loads is in the right direction. The utilities cannot completely control this balancing.

Electronics has evolved in the presence of utility power. Early circuits used vacuum tubes which meant power transformers and relatively high voltages. Any engineer designing hardware learned very quickly that shields were needed around signal conductors or that the coupling to electric fields from utility power will be a problem. It was noted that if the mounting hardware is connected to earth, the coupling is reduced. This in turn led to the idea of providing a dedicated ground to be used for signals, shields, and racks containing electronics. This idea of a single-point ground for all electronics became law in defense installations. The code was changed to allow *equipment grounding conductors* (green wires) to be *ungrounded* (isolated) at the receptacle and *grounded* at the *service entrance.* There are many installations where this grounding procedure has introduced significant noise problems. The practice is no longer allowed in hospitals as there is an added risk rather than an added safety factor. Obviously, there are many wiring methods that are safe. Not all of the methods are compatible with noise control. The role of engineering is to choose the method that fits the need.

Efforts to provide separate *equipment grounds* for parts of a facility are unsafe as some faults could go undetected.[1] The code will permit floating power but only when there is a specific need and where there can be close supervision. Examples are some assembly lines and electrically heated crucibles. A fault detection system must be provided.

The word ground has many meanings to engineers. It has many nontechnical meanings as well. It is a word that seems to attract many uses. The National Electrical Code provides definitions that are quite clear. This code is law in most cities. For example, the *grounded* conductor is the current-carrying power conductor that is connected to earth at a *service entrance.* It is colored white. *Equipment grounding conductor*s cannot be used for carrying power current and they are colored green or it can be a bare copper wire. No other power conductor can be colored green. Any conductor that could come in touch with a power conductor is called an *equipment ground.* This includes racks, metal cabinets, motors, generators, and power transformers. All *equipment grounds* are connected together and are earthed at the *service entrance.* Any electrical hardware that uses utility power that could be a source of electrical shock must be *grounded* per code. The code establishes grounding rules that guarantees that any fault current path will be low impedance so that a breaker will trip if there is a fault. All equipment used to carry power wiring must be *listed.* This means it has passed tests that guarantee performance over time in all types of weather. It is illegal to modify or tamper with listed hardware.

Ground becomes an issue in the desert. In the summer, the earth's surface is an insulator. It is impractical to seek a conducting layer by drilling down hundreds of feet. The same thing happens on lava beds or on granite outcroppings. An earth connection can be impractical. There can still be lightning strikes and fault conditions that can be dangerous. The general rule is that all common conductors that a human can touch should be bonded together forming a grid. There are thunder storms in the desert and this can damage distribution transformers using *isolated grounds*. Lightning can strike a jet aircraft but as long as it is one conducting surface no damage will result. Two comments: Lightning currents must not enter an area where fuel vapor can be ignited. At the base of towers, the current must be dispersed over the earth's surface so that there are no points of concentration in the structure. Concentration of current can melt metal and topple a structure.

---

1 Earth connections have resistances of around 20 ohms. A 120 V fault between two earth grounding points would result in a current of 6 A. This is hardly enough current to trip a breaker. A fault might allow voltage between nearby equipment grounds. Coming in contact with these two grounds could be lethal.

The earth as a conductor has conductivity that varies over a wide range. Skin effect is dominant when the surface is damp and current will flow on the surface. A nearby lightning strike will electrocute cattle standing on wet soil. The ocean is a good conductor and the current from a lightning strike will stay on the ocean surface. This is skin effect. A strike to a ship will not enter the water much below the water line. This has implications in providing current paths in lightning protection. For example, on a boat, current will not use the keel to enter the water. It may arc through an insulated hull at the water line.

In circuit board designs the common or reference conductor at 0 V is often called ground. In many circuits there is no earth connection. Obvious cases are cell phones and installations in automobiles or aircraft and of course, satellites.

4.9 Grounds on Circuit Boards

The ground plane used in a logic structure is a convenient way to form transmission lines. A separate return conductor for each transmission line would also function but it takes more etching and this reduces mechanical stability. Engineers often try to make measurements of voltage drops along a transmission line. The voltage at the ends of the transmission line is available on pads but the voltage at midpoints on traces requires cutting through any conformal coat or dielectric. There are several comments that might be helpful. A voltage measurement is really a field measure and the field being measured can originate from nearby sources. The presence of a probe can change the amplitude of the field so that the measure may be invalid. Here are two precautions to take. Connect the tip of the probe to the probe common. Make sure there is no signal when the probe is near the board. Connect the shorted probe to the board common to make sure there is no signal. If there is signal there may be current flow in the probe shield that is coupling field into the input. This would indicate that a better probe is needed.

A question often arises as to how to connect a circuit board to a conducting enclosure. If the circuit board receives data over a cable then an added ground might allow fields a second transmission line path that uses the ground plane of the circuit board. This is an example of where a single-point ground is preferred. The board should be grounded so that any field that uses this added transmission line does not use the board ground plane as one of its conductors. If interfering fields terminate on the ground plane it could add interference. Stated another way, the ground connection should not form a current path so that interference currents flow on the ground plane.

INSIGHT

The field that causes current flow is inside a conductor. Any loop that is formed to make a measurement of ground potential difference will probably couple to other nearby fields negating the validity of the measurement.

4.10 Equipment Ground

Hardware that is rack mounted usually makes use of utility power. The National Electrical Code requires that the rack and all hardware mounted to it be treated as equipment and *grounded* per code. Power line filters use the *equipment ground* as a return conductor. The current flow can result from activity in the hardware or coupled to the power conductors from other hardware. In large systems, these line filters can place a significant reactive load on the line. The fields associated with this filtering are not totally confined and can contribute to the general ambient field in a facility. The ambient noise field intensity will be reduced if this current is provided many parallel paths to follow. This is the reason that *isolated equipment grounds* are not effective. An example will illustrate the problem. Two computers are coupled together. Each is powered using an isolated ground. A nearby elevator motor creates a transient field that couples field energy to the equipment ground loop. The logic connection is closure for the equipment ground loop and the induced voltage transient blows up circuitry at the interface. Rather than reduce interference, the *isolated grounds* created problems. If the computers were coupled using an rf link or fiber optics, this would not occur. The less costly solution is to connect all *equipment gr*ounds together in a grid.

4.11 Guard Shields

In analog instrumentation, signals are often in the millivolts. To maintain signal integrity, all errors of 0.1% or greater must be considered. There are literally dozens of mechanisms that can contribute errors at this level. One major problem is the ground difference of potential between the signal source common and the instrumentation output common. This potential difference can often exceed 10 V. For a 10-mV signal with a source impedance of 1000 ohms, the error voltage is limited to 10 μV. For a 10-V common-mode signal, the error current level is limited to 10^{-8} A. This is an input impedance of 1000 megohms. At 60 Hz this is a leakage capacitance of 2.7 pF. This is a difficult restraint on both the signal cable and on the instrument design. These figures are presented to show why it is desirable to avoid this approach to analog signal processing.

The potential difference between grounds is a measure of the field intensity in the intervening space between a source transducer and any signal processing. The ground difference in potential is called common-mode interference. The arriving shield is called a guard shield and it guards the signal conductors up to the input integrated circuit pads. This shield is grounded at the transducer and not connected to the local common. A simplified schematic of this circuit arrangement is shown in Figure 4.6.

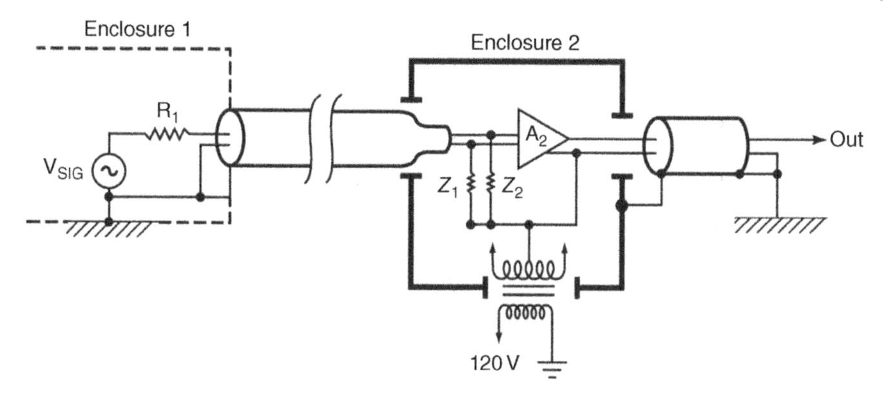

Figure 4.6 A differential amplifier using a guard shield.

Filters and clamping circuits protect the input against any overvoltage that might occur. The guard shield is not expected to be effective above 100 kHz. The guard shield is connected to local common above this frequency. A typical connection is a 0.01 μF capacitor in series with 100 ohms. If the input shield were to be grounded to the local common there would be a voltage gradient along the shield and this would couple interference directly into the signal path. Transient common-mode signals can often exceed 100 V. In practical applications, rejection ratios must exceed a million to one at frequencies of 100 Hz.[2] This requires the amplifier input impedance Z_1 and Z_2 must be 1000 megohms. This problem is avoided if the signal is amplified, digitized, and carried by fields to a remote point for processing and data storage. The field can use free space or fiber optics but not a hard wire connection.

4.12 Forward Referencing Amplifiers

Analog signals that are preconditioned must often be brought to a logic board for processing. The signals are often carried on a single coax cable. The signal level can be bipolar and limited to ±5 V. The source impedance is usually under 1 ohm and the bandwidth can be as high as 100 kHz. The basic problem is that the reference conductor is at a different potential than the logic board reference conductor. Connecting the two reference conductors together invites current flow that can result in interference. The interference can impact the analog signal and in some instances it can impact the logic.

A simple differential amplifier with unity gain can buffer the signal and limit the common current flow to a few milliamperes. The common of the analog

2 Tests that take place in isolated areas between test structures and buildings often experience conditions where ground potential differences exceed hundreds of volts.

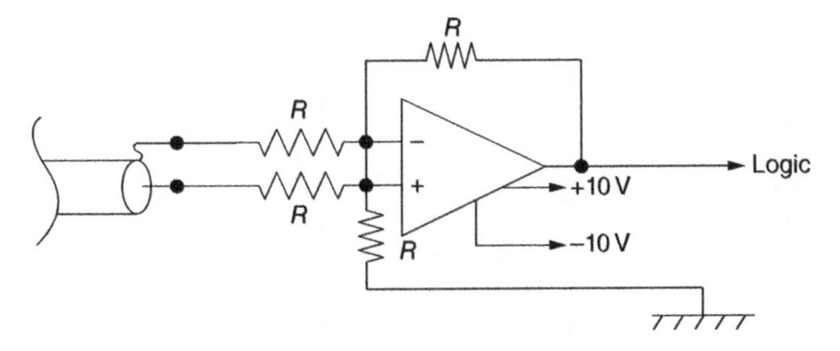

Figure 4.7 A forward referencing amplifier.

signal can be connected to either of the differential inputs depending on the desired gain polarity. The two commons are not connected together. The circuit arrangement is shown in Figure 4.7. This circuit would normally require a negative power supply voltage. This circuit is not intended for terminating a transmission line. The use of 50-ohm feedback resistors would require current levels of 100 mA. Thousand-ohm resistors are usually very effective. Typically, analog IC amplifiers have limited current capability.

4.13 A/D Converters

The interface between an analog signal and an A/D converter on a logic board presents a unique problem. The analog signal is referenced to a remote source ground. The A/D converter is referenced to logic common. The input to the A/D converter is usually a forward reference amplifier similar to the one used in Figure 4.5.

An A/D converter might have 12-bit resolution which means that the interference level must be below 1 mV. It is important not to couple normal-mode interference to the signal path. If the signal common is grounded, current flowing in this conductor can easily add normal-mode interference that cannot be rejected by the forward referencing amplifier. The best practice is to separate analog signal space from logic signal space. There can be only one common conductor. The analog space must use dedicated pins, traces, and power connections. The analog input signal loop must have minimum area. Any attempt to use separate grounds will only increase the interference coupling.

4.14 Utility Transformers and Interference

Utility power plays an important role in moving both energy and interference throughout a facility. In industry, utility power runs computers, lights, elevators, assembly lines, and communications systems. There are many opportunities for interference in the form of field generation and transmission through the utility lines. Transformers in individual devices allow the power source to be referenced to many different common points. The transformers in each device provide a path for interference to travel in both directions. In many products, energy is transferred using dc-to-dc converters which can add a pulse-like character to the interference.

The electrical code allows for only one grounding grid in a facility. The code allows for *separately derived* sources of power and for separate distribution transformers, but for safety there can be only one grounding system. Power supplied from a backup generator or from an added transformer is said to be *separately derived.* The *neutral* and all *equipment grounds* are brought back to each power source and *earthed* to the nearest available point. The important point is that power from a *separately derived* transformer allows the use of a dedicated *neutral.* The voltage drops in a shared *neutral* can often be a source of interference. The *neutral* or *grounded* power *conductor* can connect to earth and the *equipment ground* only once and that is at the *service entrance.* Each source of power is treated as a *service entrance.*

Utility power is usually generated three phase. The coils that rotate in a magnetic field that generate the voltage are spaced 120° apart. The reason for three-phase power relates to the torque on the rotor. If the phase loads are balanced the torque on the rotor is constant over every revolution. Without a fixed torque, vibrations would tear up the shaft bearings in a very short period of time.

Distribution transformers for a building are sometimes centrally located to reduce cost. The leakage magnetic field near the transformer can be a source of interference. This field is proportional to secondary load current that can be quite high. In installations of this type the limit of magnetic field strength at full load should be specified. Attempts to shield this type of interference are usually not successful.

Distribution transformers should not be mounted where leakage fields can couple current into the building steel. Nearby conducting loops should be broken by using insulators.

4.15 Shielding of Distribution Power Transformers

In facilities where power is used for industrial activities, there can be electrical interference carried on the power conductors. There are many levels to interference including power surges, voltage sags, spikes, and high-frequency transmissions. Power conductors are transmission lines that can carry interference for long distances. In critical installations, a shielded distribution transformer can be used to service specific hardware. Internal to the distribution transformer there can be three shields. These shields consist of three conducting layers of either copper or aluminum that surround the primary coils of the transformer. The shields must not form shorted turns around the iron core. The purpose of the shields is to control the paths taken by interfering reactive currents. The primary coil is usually wound next to the core.

The first or primary shield is placed over the primary coil. It is connected to the end coil of the primary winding. The center shield is connected to the equipment ground or metal frame around the transformer. The third shield is connected to the start of the secondary coil.

The center shield is the common-mode shield. It blocks the flow of common-mode currents on the primary conductors from flowing in the secondary circuit through coil-to-coil capacitance. The primary and secondary shields keep common-mode currents from flowing in leakage capacitance to the coils of the transformer which can then enter the secondary circuits by transformer action.

In a three-phase transformer, each of the three legs are separately shielded. All shields are connected inside the transformer. This shielding is shown in Figure 4.8.

4.16 Electrostatic Discharge

There are many situations that can cause charges to accumulate on an insulated body. The friction of shoes on a rug or on floor tiles can build up a charge on a human body. Charges can build up on an insulated chute carrying moving grain. The discharge of this energy can start fires or cause explosions. An aircraft in flight can build up a charge moving through air. Fueling of aircraft requires careful grounding procedures so that arcing does not occur around fueling. Charges can build up on an insulated bag when integrated circuits are removed. This arcing can damage semiconductors. If a discharge takes place on a cell phone or on a keypad, there must not be damage to circuitry. Manufacturers must test their hardware to make sure that it can survive ESD pulses. In the manufacture of integrated circuits, humidity is controlled and there are clothing restrictions. Low-voltage arcing can produce problems that can go undetected. In computer installations, floor tiles are made slightly conductive so that any charge buildup is immediately discharged. This means that each tile must be correctly grounded. Grounding clips must be replaced if removed. Rotating floor polishers are not allowed.

Humidity control in a facility is the best way to limit ESD problems. Adding humidifiers is not always effective as there can be areas that do not get treated air.

All products that contain electronics should be tested to make sure they are not damaged by ESD. Performance in the presence of ESD is usually not expected. Any external conductors that provide an electrical path into the

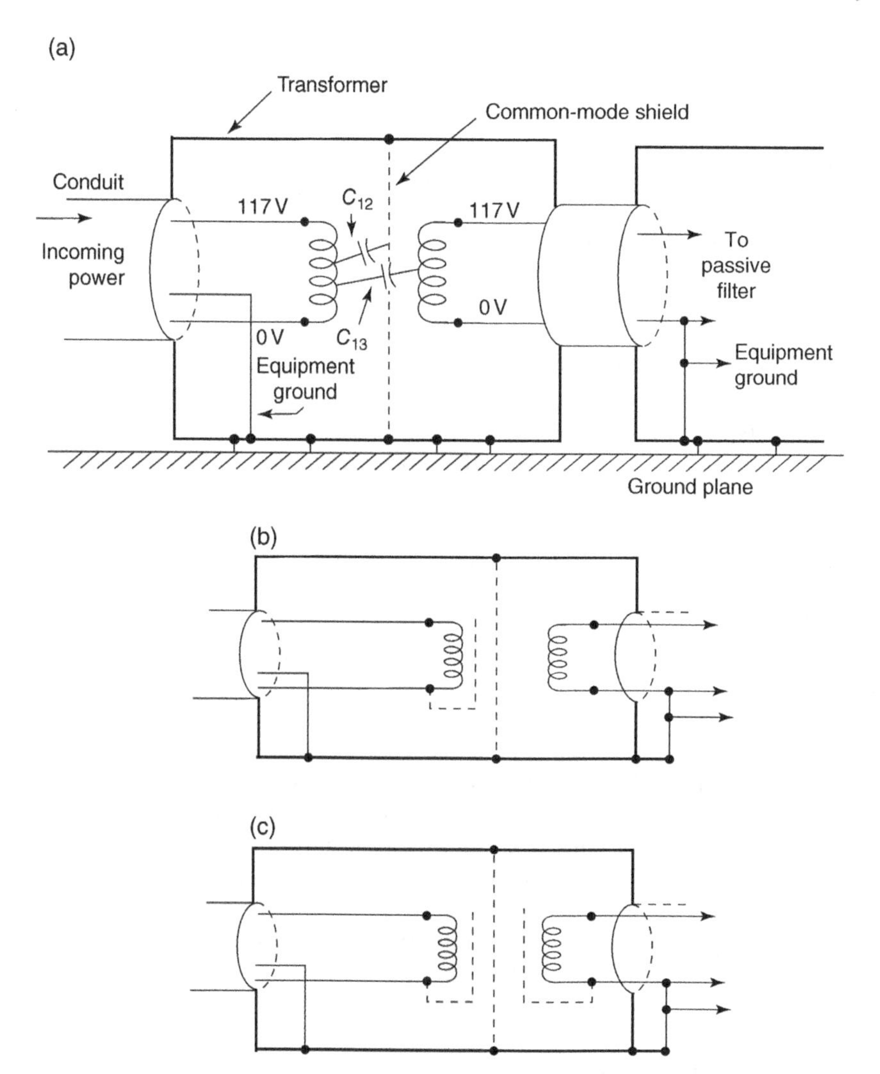

Figure 4.8 A single-phase isolation transformer. (a) One shield, (b) two shields, and (c) three shields.

electronics should be tested. This includes power and signal cables. Devices that are grounded through a power cord are more vulnerable than devices that are battery operated. The issue is: Where will the arcing take place and what is the current path? These two factors define where the interference field is located and what the field intensity will be at points of difficulty. Energy can enter through apertures, seams, or pins on open connectors. Energy can enter on shields that are not properly terminated, power leads that are not properly filtered, and through displays. A control that is mounted on a painted or anodized surface can provide an ESD entrance.

An ESD current pulse creates a significant magnetic field that can couple into any nearby loop area. If the current spreads out over a conductive surface, the field intensity is significantly reduced. Providing multiple current paths for the current is obviously in the right direction. Single-point connections concentrate current flow and this in turn increases field intensity.

Commercial zappers are available for testing hardware that generate controlled pulses of current. The voltage that initiates the pulse can be varied from 1,000 to 15,000 V. The current level is typically 5 A. An ESD pulse can be characterized as being a sine wave current at a frequency of 300 MHz. The pulses can be single or repeated events. There are two basic modes of testing. One mode is where the probe makes ohmic contact and there is no probe tip arcing and the second mode is when the probe arcs to the device under test.

An example of coupling provides some insight into the nature of the problem. Consider an ESD pulse of 5 A. Using Equation 1.16, the H field at a distance of 10 cm is $I/2\pi r = 79$ A/m. Using Equation 1.18, $B = \mu_0 H = 10^{-4}$ teslas. Consider a loop of 1.0 cm^2 coupling to this field. Using Equation 1.19 at a frequency of 300 MHz, the voltage induced in this loop is 19 V. This voltage will destroy an integrated circuit. The current pulse does not need to flow in a component to do damage.

A problem exists when the device being tested is not grounded. Repeated zapping can build up a stored charge that may eventually cause a breakdown that would not normally occur. If this condition exists, a discharge path using a 100-megohm resistor can be added. Devices that use a power adapter can be vulnerable when the voltage buildup can arc across the power coils in a transformer. The buildup of static charge limits the voltage difference to the zapper and this can invalidate the test.

Testing should proceed at selected intermediate voltages. High voltage arcs lose a lot of energy to heat. The most critical voltage is often at about 7000 V.

4.17 Aliasing Errors

An example of an aliasing error occurs in old Western motion pictures where wagon wheels appear to rotate backward. To avoid this error the data must be sampled and displayed at a higher rate. The rule is that data should be filtered and then sampled more than two and a half times the highest frequency of interest. If the data is temperature, it is easy to assume that sampling at a high frequency is unnecessary. The problem that exists is that there may be interference mixed with the signal and sampling this interference can result in a measurement that looks like temperature. The accepted way to avoid this problem is to filter the data before sampling. Filtering should not be done digitally after sampling as there may be aliasing errors. Filtering data in the analog domain is expensive but often it is necessary.

If the sampling rate of the wagon wheels is slowed, the reverse wheel rotation rate will appear to increase. This apparent shift in spectrum can be used as a test to determine if an aliasing error has occurred. With the enormous capacity we have for storing and processing data, a simple spectral analysis is very practical. The data is analyzed at two noninteger sample rates. If the spectrum shifts linearly the data is invalid.

Glossary

Aliasing error A sampling error that results when data contains frequency content close to the sampling rate (Section 4.17).

Cross-coupling The unwanted transfer of energy between circuits. This is often the energy transferred between transmission lines. Long transmission lines are the most vulnerable. The victim wave travels in the opposite direction to the culprit wave.

Equipment ground All conductors that could contact a utility power conductor. These conductors include racks, cabinets, housings, junction boxes, conduit, raceways, generators, motors, breakers, and so forth. Equipment ground conductors are green or solid wires running next to current-carrying conductors to keep the fault paths low impedance. They connect to the neutral conductor at one point in the service entrance. Equipment grounds can be earthed at multiple points.

Electrostatic discharge (ESD) The high-voltage phenomena where friction moves charges so that they accumulate on bodies and arc when the breakdown potential is reached (Section 4.16).

Even-mode coupling Coupling that adds to a logic signal and impacts signal integrity (Section 4.7).

Forward referencing amplifier A unity gain differential amplifier used to buffer analog signals. The input common can connect to either input thus controlling gain polarity. The amplifier is used to couple signals referenced to ground in one electrical environment to ground in a second electrical environment. The principle use is to avoid a ground loop. The common-mode rejection ratio is usually about 100:1 (Section 4.12).

Grounded conductor In single-phase power this is the return current conductor.

Guard shield A shield that surrounds the conductors carrying a single analog signal that is connected to signal common at the signal source that is usually grounded at the signal source (Section 4.11).

Impedance of space The ratio between the E and H fields for electromagnetic energy traveling in space at a distance from a radiator (Section 4.3).

NEC National Electrical Code.

Neutral The return power conductor that is earthed at the service entrance in three-phase power. This conductor can carry power current. In a balanced three-phase circuit the neutral current is zero (Section 4.8).

Rise time The time it takes for a voltage to rise to its expected level for step or square waves. Typically when exponentials are involved, the rise time is measured between 10 and 90% points. This is not a standard. Rise time is a guide to determine the bandwidth of a system.

Zapper An instrument for generating pulses that simulate the effects of ESD (Section 4.16).

5

Radiation

5.1 Introduction

The number of circuit board applications that involve transmitting or radiating signals is growing steadily. Radio, television, and garage door openers held center stage for a long while. Today, global positioning, cordless phones, cell phones, remote monitors, automobile keys, and communications links of all types are taking over. Most devices require a circuit board that receives and/or transmits data. Many devices are battery operated, so efficiency is often an issue. It is interesting that the same technology that allows the rapid processing of digital data can also transmit and receive data in the form of modulated carriers.

The general radiation problem involves sine wave signals, active circuits, transmission lines, multiple ports, and antennas. A practical approach to handling rf design is to measure the reflection and transmission coefficients at each port. These coefficients are known as scattering or S parameters. This approach is practical because these parameters can be acquired without using short and open circuits. Just as a reminder, we are leaving the world of step functions and entering the world of sine waves. In most cases, we deal with single-frequency phenomena where the language of circuit theory is appropriate.

On a circuit board, an antenna must couple energy into free space which has a characteristic impedance 377 ohms. A driver usually connects to a transmission line with a characteristic impedance of 50 ohms. At the antenna, the impedance might be complex and near 100 ohms. To optimize the flow of energy there needs to be a matching network placed between the transmission line supplying energy and the antenna. This network plus the antenna must look like a 50-ohm termination so that there is no reflection at this junction. In applications where the radiated frequency is varied, this matching problem becomes a compromise.

Fast Circuit Boards: Energy Management, First Edition. Ralph Morrison.

5.2 Standing Wave Ratio

On terminated transmission lines, when there are reflected sine waves, they combine with the incident waves to form a standing voltage pattern. There are fixed points along a transmission line where the voltage peaks and other fixed points where the voltage never exceeds a minimum value. This pattern is static and depends on frequency and line length. For long lines, there are multiple peaks and valleys. The ratio of maximum to minimum voltage along the line is called the standing wave ratio (SWR). This ratio is used as a guide as to just how closely a transmission line is matched to its termination. When large amounts of power are involved or where there is limited power available, this ratio is important. In an ideal case, the ratio is unity. Where the voltages peak, the currents are a minimum. This means that where the electric field energy peaks, the magnetic field energy is minimum. A pattern of superposed waves is shown in Figure 5.1.

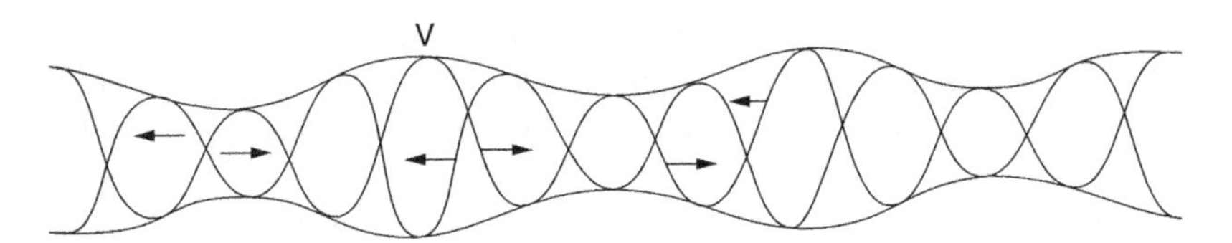

Figure 5.1 A standing wave pattern.

> **INSIGHT**
>
> In Chapters 1 and 2, it was pointed out that energy is not reflected at a transmission line transition. Wave action is the process of storing, converting, or moving magnetic and electric field energy. The common waveforms are step functions or sinusoids. For step functions, it was easy to identify when there was static energy and moving energy. For sine waves these transitions still take place but on a continuous basis which makes it very difficult to present graphically on a printed page. There are presentations on the internet that display moving sine waves, but they do not display both electric and magnetic field intensities.
>
> For sine waves, all parts of the wave are acting as a leading edge. Standing waves are an indication of where along the line the electric and magnetic field energy storage tends to concentrate. When there are no standing waves, energy passing every point is sinusoidal at one amplitude and moving forward.

5.3 The Transmission Coefficient τ

In Section 2.10, the transmission coefficient τ was introduced. In this discussion all impedances were resistances or real numbers, so τ was a real number.

$$\tau = \frac{2Z_L}{Z_0 + Z_L}. \tag{5.1}$$

This equation is valid for real and complex terminating impedances. As an example, if a capacitor terminates a transmission line, the impedance Z_L is a reactance of $-j/\omega C$. If $C = 2$ and $\omega = 1$, then $Z_L = -0.5j$ ohms. If $Z_0 = 1$ then using Equation 5.1

$$\tau = \frac{-1.0j}{1 - 0.5j} = 0.4 - 0.8j. \tag{5.2}$$

If the forward wave is a 5-V sine wave at $\omega = 1$, the transmitted wave is $5\tau = (2 - 4j)$ V.

For a review of circuit theory and complex numbers see Appendix A.

5.4 The Smith Chart

The Smith chart can be used as a work sheet for designing matching networks to terminate a transmission line in its matching impedance. The chart deals with sine waves at one frequency. The chart treats the S parameter problem of one port and one reflection coefficient. Working charts are available on the internet. A Smith chart with limited resolution is shown in Figure 5.2.

A Smith chart[1] is a plot of the complex transmission coefficient τ for a transmission line with a characteristic impedance of 1 ohm terminated in any impedance Z_L. Assume $Z_L = r + jx$. The point Z_L on the chart is at the intersection of circle r and circle x. The value of τ is the distance from the point (0,0) to the intersection of circles r and x. The Smith chart works for any transmission line length at any frequency. An examination of the chart shows that x and r can be any value from 0 to ∞. Note that x can be of either polarity. The value of τ is not shown on the graph as it is not needed for most applications.

The center or origin of the Smith chart is where $\tau = 1$ and there is no reflection. The distance from the point (1,0) to Z_L. is $(\tau - 1)$. Since

$$\tau = \frac{2Z_L}{Z_0 + Z_L}, \tag{5.3}$$

1 A paper authored by P.H. Smith titled "Transmission Line Calculator" described this chart in Electronics—January 1939.

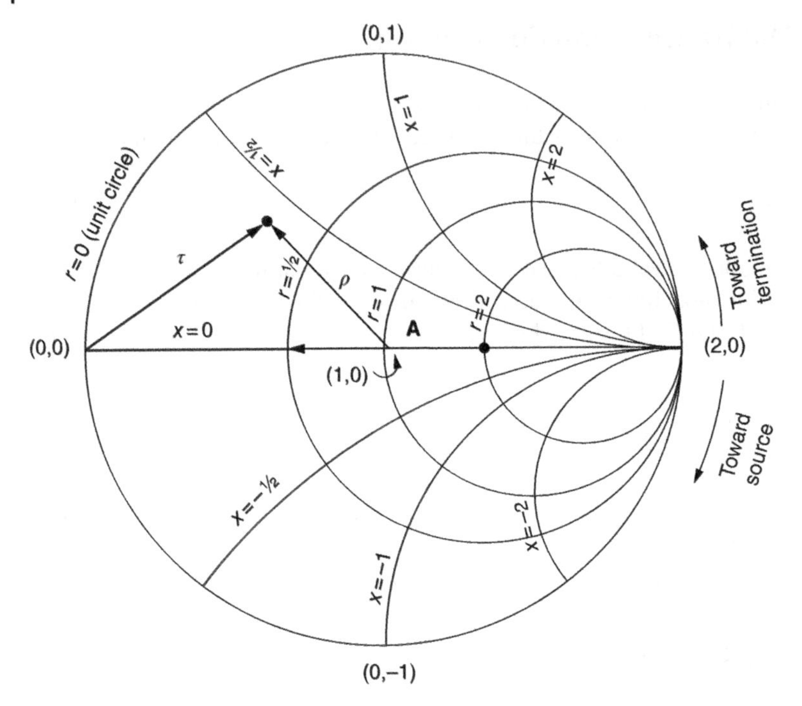

Figure 5.2 An impedance Smith chart showing the relation between the reflection coefficients and terminations on a 1-ohm transmission line.

then

$$(\tau - 1) = \frac{Z_L - Z_0}{Z_L + Z_0} = \rho. \tag{5.4}$$

Thus, the distance from the point (1,0) to the terminating impedance Z_L is the complex reflection coefficient ρ. This equation was introduced in Section 2.10.

The circles on the Smith chart are marked r and x where r stands for real value and x stands for imaginary value. This makes it possible to interpret the curves to mean a value of impedance or admittance. For impedances, x stands for reactance and r stands for resistance. For admittances, x stands for susceptance and r stands for conductance. The user must decide when the chart represents impedance and when it represents admittance. For an impedance chart, the point $Z_L = 0$ is where x and r are equal to 0. This is where the terminating impedance is a short circuit. The point at the far right, where x and r are infinite, is an open circuit. If the chart represents admittance, then the point where x and r are 0 represents an open circuit and the point where x and r are infinite represents a short circuit. To use the chart for admittance the value of Z_L can be reflected about the origin. A resistor of 5 ohms is on the circle $r = 5$. A conductance of 0.2 S is on the circle $r = 0.2$. A reactance of $2j$ is on a circle of $x = 2$. A susceptance of $-0.5j$ siemens is on a circle of $x = -0.5$.

In a typical design, it is desirable to carry energy on a 50-ohm line to an antenna structure and match its driving point impedance. To work with a Smith chart, the antenna impedance needs to be scaled down by a factor of 50. The Smith chart can then be used to change the terminating impedance to 1 ohm. This is done by adding impedances and admittances to the antenna at the interface. The added elements are then scaled up in impedance by a factor of 50. This matching of impedance only works at one frequency.

If the Smith chart represents impedances, then moving along a circle of constant resistance adds or subtracts reactance. Adding reactance means adding a series inductor or reducing an existing capacitor. Subtracting reactance means adding a series capacitor or reducing an existing inductor. Moving along a circle of constant reactance adds or subtracts resistance. A negative resistance is not allowed. If there is a resistor present it can be reduced.

If the Smith chart represents admittances then moving along a circle of constant conductance adds or subtracts susceptance. Adding susceptance means adding more shunt capacitance. Subtracting susceptance means adding a shunt inductance or reducing an existing capacitor. Moving along a circle of constant susceptance adds or subtracts conductance. Adding conductance means reducing the value of a shunt resistor.

The technique used to match impedances usually does not add resistance or conductance as these components dissipate energy. This means that the correcting impedance is usually built up of inductors and capacitors. The idea in construction is to add or subtract impedances or admittances until the transmission line has an input impedance (admittance) of one real ohm (one real siemen). The only orthogonal circle that crosses the (1,0) coordinate is $r = 1$. This real circle is thus key to designing a terminating network. The usual procedure is to add reactances or susceptances to reach this circle. This makes it easy to reach the point (1,0) by adding the last reactance or susceptance.

To illustrate the use of the Smith chart, consider a termination of 100 ohms on a 50-ohm line. On a Smith chart this is a termination of 2 ohms. This impedance must be modified to where the input impedance is 1 ohm. We rule out paralleling the 2 ohms with a resistor of 2 ohms as half of the energy will be lost. Since parallel elements must be used to reduce the impedance, the admittance form of the chart is used first. The reciprocal of 2 ohms is 0.5 S. This point is located diametrically opposite the midpoint of the chart. This move is shown as step A on Figure 5.2. The chart is now in admittance mode. We need to place a shunt capacitor across this conductance that will add to the susceptance to reach the dotted admittance circle. The path B is a distance of $0.5j$ siemens along the $r = 0.5$ circle. This move represents a shunt capacitor with a reactance of $-2j$ ohms.

We now convert back to an impedance Smith chart. The reciprocal of the point $0.5 + 0.5j$ is the point 1.0 and $-1.0j$ on the path C. If an inductor of $1.0j$ is added, the resulting impedance will be $1 + 0j$ which is our objective. In review, we now have an impedance of 2 ohms in parallel with a reactance of $-2.0j$ ohms in series with a reactance of $1.0j$. The transmission line is now terminated so that the input impedance is 1 ohm.

These reactances can be related to capacitors and inductors once a frequency is selected. At 500 MHz for a 50-ohm line the inductance is 159 nH and the capacitor is 6.7 pF.

This problem can be solved using circuit analysis. The impedance of an inductor in series with an admittance consisting of 2 ohms in parallel with a capacitor C is

$$Z = j\omega L + \frac{1}{2 + j\omega C} = 1.0 \tag{5.5}$$

The real and imaginary terms must balance separately. It is easy to show that $\omega C = ½$ and $\omega L = 1$ which agrees with the Smith chart procedure.

A second solution is practical where the capacitor and inductor are interchanged but with different values. This solution is shown in Figure 5.3b. Step A is the same as before where the chart is in the admittance form. Step B shunts the line with an inductor to reach the mirror image of the $r = 1$ circle at $x = -1/2j$. Step C returns the chart to the impedance form. Step D adds a series capacitor to reach the $(1 + 0j)$ point. Again, the input impedance to the transmission line is 1 ohm.

When parasitics are included in the problem, the mathematics can become complex and a graphical solution can be very useful. In general, accuracy is not an issue in designs as there are many parasitics that are not considered. The Smith chart provides a visual tool that directs finding the path to a solution.

5.5 Smith Chart and Wave Impedances (Sine Waves)

A transmission line terminated in its characteristic impedance has an input impedance that is the characteristic impedance of the line. This impedance is independent of line length. If a transmission line is terminated in any other impedance, the input impedance varies with line length and frequency. The input impedance varies because the reflected wave modifies how much current the source must supply. The line length determines the phase of the reflected voltage at the source and this affects the input impedance.

On a Smith chart the reflection coefficient vector $\boldsymbol{\rho}$ is drawn from the center of the graph to the terminating impedance. The terminating impedance is the wave impedance of the line at the termination point. To determine the

(a)

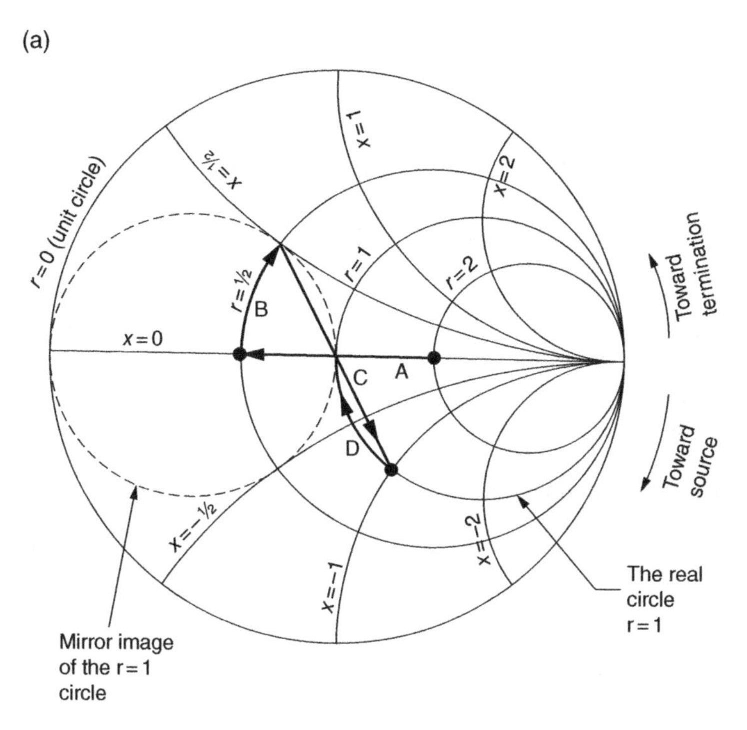

(b)

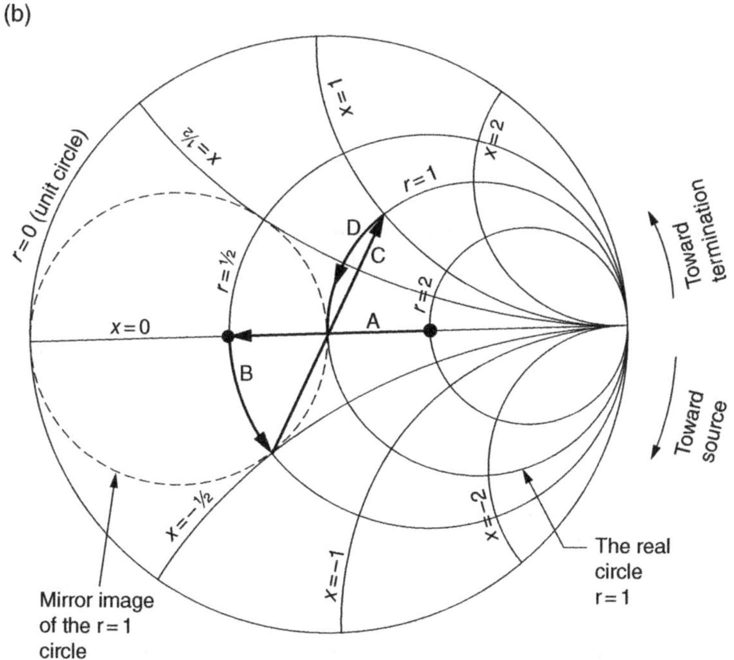

Figure 5.3 (a) The paths taken on a Smith chart to reach the point $\tau=1$ where $x=0$ and $r=1$. (We first added a shunt capacitor.) and (b) The paths taken on a Smith chart to reach the point $\tau=1$ where $x=0$ and $r=1$. (We first added a shunt inductor.)

impedance at the driving point, the reflection vector can be rotated back (clockwise) toward the source. The angle of rotation is a measure of length on the transmission line. One half wavelength corresponds to 360° on the Smith chart. The reflection coefficient vector points to an impedance that is the input impedance of the transmission line for that line length and for that termination.

To determine the SWR along a line, the reflection coefficient vector is rotated to where it crosses the real positive axis of the chart. The crossing point is the SWR. The proof is as follows.

The SWR is the maximum voltage on a transmission line divided by the minimum voltage. The maximum voltage occurs where the forward and reflected waves add together. The minimum occurs when they subtract. The SWR at an arbitrary point on the line is

$$\mathrm{SWR} = \frac{V_F + V_R}{V_F - V_R} \tag{5.6}$$

where V_F is the amplitude of the forward wave and V_R is the amplitude of the reflected wave. The reflection coefficient $\boldsymbol{\rho}$ relates the forward wave to the reflected wave or

$$V_R = \rho V_F. \tag{5.7}$$

Substituting Equation 5.7 into Equation 5.6 yields

$$\mathrm{SWR} = \frac{1+\rho}{1-\rho}. \tag{5.8}$$

On the Smith chart the reflection coefficient $\boldsymbol{\rho}$ is the vector distance from the point $Z = 1$ real ohm to the terminating impedance Z_L.

The reflection coefficient is given by

$$\rho = \frac{Z_L - Z_0}{Z_L + Z_0}. \tag{5.9}$$

Solving for Z_L where $Z_0 = 1$ we get

$$Z_L = \frac{1+\rho}{1-\rho}. \tag{5.10}$$

Thus, on the Smith chart, the value of the terminating impedance is the value of the SWR at that point. At a general point on the chart, the ratio of voltages is complex. The ratio of voltages we are interested in is a real number. This real number is obtained by rotating the vector $\boldsymbol{\rho}$ to where it crosses the positive real axis of the Smith chart. The angle that is rotated defines the location on the transmission line where the largest voltage ratio occurs. This real value of $\boldsymbol{\rho}$ is the SWR.

The peaks of sine wave voltage along a transmission line repeat every half wavelength. The Smith chart provides information for each one-half wavelength section. If the line length is not a multiple of a half wavelength, the Smith chart applies to the partial section up to the generator (voltage source). The reflection vector $\boldsymbol{\rho}$ points to the load impedance Z_L at the end of the line. If the vector $\boldsymbol{\rho}$ is rotated clockwise the number of electrical degrees on the transmission line, it will point to the input impedance of the transmission line when the line is terminated in Z_L. Remember that each 360° rotation of $\boldsymbol{\rho}$ represents a half wavelength along the line.

5.6 Stubs and Impedance Matching

Stubs can often be used as a reactance to match a transmission line to a load. The reactance of a stub depends on the length of the stub and whether the stub is terminated in a short circuit or left open (unterminated). Since there are no resistors involved in the termination, the stub can only be reactive which means it looks like a single capacitor or inductor. The matching process simply adds a stub to the transmission line at a calculated point on the transmission line. The characteristic impedance and the length of the stub determines the reactance that is added.

The reactance of a stub can be determined by using a second Smith chart. In the impedance form of the chart, the point where the termination is a short circuit is at the far left. The reflection coefficient is a vector from the origin to this point. When this vector is rotated counterclockwise, it points to the reactance of a shorted line where the angle is proportional to stub length. One revolution of the vector corresponds to one-half wavelength at the frequency of interest. The vector first points to a negative reactance and zero resistance. As the angle increases the point on the stub length increases and the negative reactance increases. At 180° the line looks like an infinite capacitance or a short circuit. In the next 180° the reactance is positive which means the stub looks like a decreasing inductive reactance. Thus, a shorted stub can take on any value of reactance depending on stub length.

A Smith chart can be used to determine the stub length required to provide a required reactance when the stub is open circuited. The reflection coefficient vector points from the origin to the far right. As the vector rotates counterclockwise the positive reactance decreases which means the inductance is decreasing. At an angle of 180° the inductance is 0. As the angle increases the reactance goes negative which means the stub looks capacitive. The stub can take on any reactive value depending on stub length. In general, a shorted stub is preferred over an open stub as it is less apt to radiate.

To match a transmission line to a termination, a stub is added to the line at a point that has a real normalized resistance of 1 ohm. The reflection coefficient is rotated from the terminating impedance point until it crosses the $r = 1$ circle. The angle of rotation defines the point on the line where the stub must be attached. In general, it can be located at one of two points. If the reflection coefficient crosses the 1-ohm circle at $2j$, the stub must have an input impedance of $-2j$. With this correction, the input impedance to the transmission line is 1 ohm.

Electrical degrees around a Smith chart and distance along a transmission line are proportional. At 100 MHz in a vacuum, one wavelength is 30 cm. If the dielectric constant is 4 the distance is 15 cm. A half wavelength is 7.5 cm. This is 360° on the Smith chart. Thus, 45° represents 0.93 cm along the transmission line.

5.7 Radiation—General Comments

Transmission lines by their very nature confine electromagnetic fields. We have discussed cases where reflections take place where there are transitions in conductor geometry. When sine waves are used on transmission lines, the same rules apply. In logic, reflections at open circuits are desired. For sine wave transmissions, reflections are undesirable as they imply an SWR and this limits the antenna radiation.

There are two basic radiator geometries on circuit boards. These are loops and isolated conductors that should be correctly called antennas. Loops are generally considered low-impedance radiators while isolated conductors are considered high-impedance radiators. Near a loop, the magnetic field dominates and the ratio between E and H fields is low. Near a conductor where the return current is the D field in space, the electric field dominates and the ratio between E and H fields is high. At a distance from a radiating structure, the ratio E/H approaches 377 ohms, the characteristic impedance of free space. This distance is $\lambda/2\pi$ where λ is the wavelength. This transition point is referred to as the near-field/far-field interface distance.

5.8 Radiation from Dipoles

The simplest radiator is the half-dipole antenna. On the earth, the antenna is usually mounted vertically and driven from a sinusoidal voltage source at the base. The current supplied to the antenna flows in space and returns near the transmission line on the surface of the earth at the base of the antenna. In space, this current is correctly called a displacement current. In almost all cases the driving voltage is a sinusoid. The information being transmitted can modulate the amplitude, frequency, or phase of transmission. In many schemes, the modulation is a combination of phase and amplitude. If there are four amplitude levels and eight phase possibilities, a length of carrier can represent a 32-bit word.

In a typical circuit carrying sinusoids, the energy stored in an inductance or capacitor is returned to the circuit in each half cycle. Most of the energy is stored in the component or in the nearby space. When a radiator is built, the field energy is allowed to extend out into space. There is a delay associated with returning this energy to the circuit. The returning energy can be divided into two components. The component that is 90° out of phase cannot re-enter the circuit and is radiated. The higher the frequency, the more efficient the radiator can be.

The electric field dominates near a dipole. This higher field impedance makes it easier to shield. The relative field intensities at a distance from the radiator are shown in Figure 5.4. Beyond the near-field/far-field interface, the E and H field intensities fall off proportional to distance, not distance squared. The energy density falls off as the square of the distance.

The field intensity around a dipole is independent of direction and falls to zero in the vertical direction. In many applications, omnidirectional radiation is preferred. The antenna geometry is not limited to simple dipoles or loops. Right-angle structures that are asymmetrical and use conducting plates are often effective in radiating in all directions. Some antennas are in the form of paddles. The presence of nearby circuitry makes it difficult to predict the actual radiation pattern.

Vias often extend through a board where the path for current flow uses only a part of the via. This unused part of the via forms a short segment of antenna that can radiate in the GHz range. To avoid this radiation, the extra via segment can be removed by drilling. A second solution is to provide a return current path next to the via so that a radiating structure is not formed. This points out the fact that field control gets more difficult as the frequency content rises.

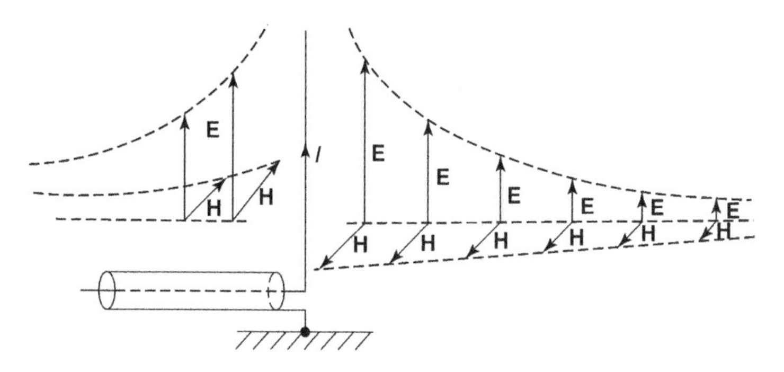

Figure 5.4 The E and H field intensities near a half-dipole antenna.

5.9 Radiation from Loops

Loops are characterized by currents that follow a closed path on conductors. The radiation from logic traces is considered a loop in character and the wave impedance is 50 ohms. Radiation is limited to that part of trace where the voltage is in transition. At a distance equal to the near-field/far-field interface, the ratio of field intensity will be 377 ohms. Inside this distance, the H field dominates. Outside of the interface distance the E and H field intensities fall off linearly with distance, not distance squared. The field intensities are shown in Figure 5.5.

The current levels on a typical circuit board are limited by the characteristic impedance of traces. The characteristic impedance of free space is 377 ohms. Dipoles are, in general, a better way to bridge this impedance mismatch if radiation is to be efficient.

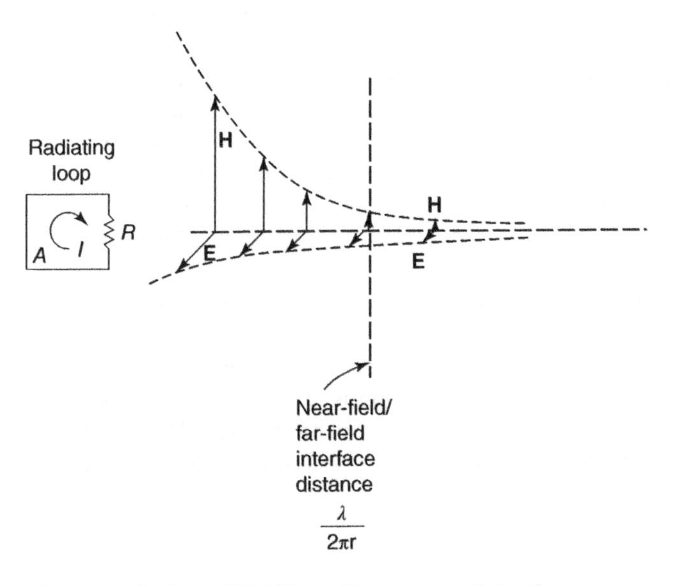

Figure 5.5 The E and H field intensities near a radiating loop.

> **INSIGHT**
>
> The concept of ground or zero potential loses its meaning in space. It is very handy to have reference conductors to describe circuit behavior. Fields are shaped and reflected by conductors, but this is not the domain of circuit theory.

5.10 Effective Radiated Power for Sinusoids

The power W crossing a small area A at a distance from a transmitting antenna is

$$W = (\mathbf{E} \times \mathbf{H})A \tag{5.11}$$

where **E** and **H** are the electric and magnetic field vectors.

At a far distance r from a radiator, the field is spherical in shape and the total radiated power crossing a spherical surface would be

$$W = \mathrm{EH}4\pi r^2. \tag{5.12}$$

In the far field the ratio of E/H is 377 ohms. Solving for H, Equation 5.12 can be written as

$$W = \frac{\mathrm{E}^2}{377}4\pi r^2. \tag{5.13}$$

Using Equation 5.8, the electric field E at a distance r can be written as

$$\mathrm{E} = \frac{(30W)^{1/2}}{r}. \tag{5.14}$$

As an example, assume a cell phone transmits ½ W uniformly in all directions. At a distance of 100 m the E field strength is 40 mV/m.

A radar pulse of 10 kW in a solid angle of 1° has an effective radiated power of 360 times 10 kW or 3.6 MW.

5.11 Apertures

Every electrical device that is housed in a metal enclosure requires seams and holes for construction, ventilation, power, and signal transport. These apertures are a path for fields to move in or out of a structure. A single conductor allows the penetration of fields at all frequencies while a hole or seam attenuates a field depending on maximum dimension. In general, the orientation of the field is unknown and the assumption that must be made is to treat the worst case scenario. If the field has a half wavelength equal to the diameter of the hole or the length of the seam, we assume the field enters or leaves unattenuated. It seems counterintuitive to consider a narrow seam as being an opening but it is the only safe interpretation. In fact, a seam is defined as an opening in a conductor where surface currents cannot freely cross the seam. The author has seen radiation from a vacuum chamber formed from two machined metal surfaces over 1 inch wide. This seam had to be closed by using a gasket that made a continuous connection across the gap. In this case, the field strength was very high and there was attenuation but not enough.

Seams formed by overlapping metal surfaces can be reduced in length by adding point contacts such as screws. The initial aperture is now a group of shorter apertures. Apertures closed in this manner are said to be dependent. The dependence arises because surface currents cannot circulate freely around each opening. The result is a group of apertures that behave as a single opening. If the attenuation factor must be 100, then 100 screws would be required to close the seam. A large screw count is not a viable solution. This is the reason why a gasket is needed as it makes hundreds of connections. Conductive surfaces such as steel or aluminum will oxidize and the metal must be plated where a gasket makes electrical contact.

Seams that form a flange can be considered a waveguide. In most situations the depth of the flange is much shorter than the length of the seam and there is very little attenuation.

Consider a hole that attenuates a field by 20 dB which is a factor of 10. A similar hole at a distance from the first hole will allow equal field penetration. This second aperture is said to be independent because currents can circulate freely around the opening. To calculate the total field penetration from independent apertures, the field intensities add directly. If the external field is 60 dBμV/m, the field inside the structure is 40 dBμV/m from each aperture. This is 100 μV for each aperture or a total field strength of 200 μV. Expressed in dB notation, the resulting intensity is 46 dBμV. Notice that we cannot add intensities using the dB notation. Obviously, this is a worst case approximation. When there are many openings, the field intensity inside an enclosure cannot be greater than the external field intensity.

An array of ventilation holes behaves as a single hole if the holes are closely spaced. A screen mesh behaves as a single opening provided the mesh is bonded at each wire crossing and the edges of the mesh are bonded to the conducting surface. Aluminum mesh is unsatisfactory because of oxidation.

Connectors, displays, and line filters form apertures that must often be sealed. Conducting gaskets are needed to close these apertures. It is necessary to close all apertures to solve an interference problem. One unfiltered conductor entering or leaving a controlled space can carry interference and negate all other attempts at control. In trouble shooting, it is advised that all fixes remain in place until a solution is found. Then the fixes are removed one at a time. This is the only way to find a solution when two or more changes are required.

5.12 Honeycomb Filters

A honeycomb filter is formed by soldering together a group of hexagonal conducting channels. The channel diameter must be less than a half wavelength at the highest frequency that is to be attenuated. At 15 GHz a half wavelength is

1.0 cm, so a honeycomb channel width of 1 cm would work. A honeycomb made of a hundred channels would form a unit that is about 10 cm in diameter. These channels are independent apertures as current can circulate freely around each opening. In this example, the attenuation factor is given by Equation 3.7 or $A = 30\,h/d$ dB where h is the depth and d is the diameter of the opening. If the ratio h/d is 3 then the attenuation factor is 90 dB. Since there are 100 apertures the resulting attenuation factor is 50 dB. This is a large attenuation factor and it tells us that very little field will enter through the honeycomb. If the interfering field is at 150 MHz there is an additional attenuation factor of 40 dB. This is the ratio of half the wavelength to the aperture opening.

The biggest problem involving honeycomb filters is the aperture that mounts the honeycomb. This aperture must be sealed with a conducting gasket that is near perfect if the waveguide is to be effective. If the mounting seal in the example above has a 1 cm long gap, the honeycomb can be totally ineffective. This means the mounting gasket must be correctly installed using a plated surface.

5.13 Shielded Enclosures

The testing of electronics for radiation requires controlled spaces. In some cases this testing can be done in a remote open area, a distance from known radiators. In other cases it is necessary to use a conductive enclosure that can limit the presence of unwanted electromagnetic activity. Scientific experimentation often requires a controlled electromagnetic environment. Supplying power to operate the hardware and test equipment becomes a part of the problem. Vents that supply air can allow the entry of unwanted fields. Lighting, physical access, and communication can also provide entry for unwanted fields. The level of interference that can be tolerated obviously determines just how difficult it is to construct the controlled space. Shielding against power-related magnetic fields is the most difficult part of the general problem. For logic hardware, interference in the power part of the spectrum is usually not a problem. We discuss the general problem as this sheds light on all the issues.

5.14 Screened Rooms

A screened room that limits magnetic field penetration at power frequencies is made from panels of silicon steel. The permeability of the steel is very nonlinear and does not provide very much attenuation for weak magnetic fields at 60 Hz and its harmonics. Since power must be used to operate test hardware and lights, some magnetic field must be brought into the room. For logic circuit boards, the emphasis is on measuring radiation above 1 MHz and low-frequency magnetic fields are usually not an issue. There are some best practices that should be followed in providing power to a screened room that adds very little cost.

Power should not be routed along the external walls of the room. The changing magnetic field from power conductors will cause current flow on the exterior walls. This occurs even if the wiring is in conduit. Some of this current crosses into the inner surfaces. Ideally, the conduit carrying power should arrive perpendicular to an exterior wall. The room should not be located near existing power panels. Multiple grounding of the screen room will provide current paths that can allow some current to cross to the inner surfaces. This means that there is field penetration. It is preferred that all electrical connections to the room be located in one area. This includes power line filters and communication links. Fiber optics is a good way to communicate to outside hardware. Any steel supporting the optics cable should be bonded to the room wall. Any hole in the wall should be extended into a metal tube so that it is a waveguide beyond cutoff. The tube can run parallel to the wall surface.

The one ground for a screened room is equipment ground. There is capacitance from the screened room to any building steel under the room. This reactive grounding path can be limited by placing the room on blocks.

Air should be provided through a honeycomb filter. The duct carrying the air should have a plastic section, so the room is not regrounded. Air filters should not require removal of the honeycomb section. Florescent lighting is electrically noisy and should not be used. A weak spot is usually the finger stock used to seal the door. If the stock is protected by a metal channel it will not catch on clothing.

5.15 Line Filters

Power line filters usually consist of series inductors and shunt capacitors. The ungrounded (hot) lead is often filtered with respect to the neutral or grounded lead. The code prohibits grounding the neutral conductor, so it must be filtered separately. All leads carried in a power connection should be filtered to the common outer surface of the conducting enclosure. In a screened room this common conductor is the outer wall. The electrical code does not allow any components to be placed in the equipment grounding conductor path. This means that equipment ground (green wire) must be connected directly to the local housing. If the equipment ground conductor enters the hardware, it will bring in interference. Electrically any filter associated with the hardware must be located external to the hardware but physically the connection can be hidden from view. This means that line filters should be supplied in a metal housing so the unit can be mounted inside of the hardware without allowing fields to enter. This mounting cannot be on a painted or anodized surface. Filters in plastic housings are very suspect.

Filtering must often cover a very wide frequency spectrum which means that several filter sections may be required. For a screened room, the line filter can be an expensive component as conductors must carry many amperes, both power leads must be filtered, and filtering must cover a wide spectrum. In designing the power entrance, the path for interference should be treated in terms of fields and the process a reflection of field energy. Power lines are not broadband transmission lines and obviously the process of reflection will not be perfect. It is wise to review the path taken by interference currents so that interference fields can be traced. If fields can enter the hardware the filter is ineffective.

The inductances used in a power filter must carry the line current. Ampere-turns can easily saturate any magnetic materials. The permeability of magnetic materials above 1 MHz is also an issue. This means that inductors used in line filters will usually be built as a solenoid in air. The parasitic capacitance of the inductors limits the high frequency filter performance. This is the reason that line filters must be built in sections. To further complicate the issue, the filter sections must be shielded from each other. All of these factors make line filters for screened rooms large and expensive. It also brings into focus the fact that many line filters are ineffective and can be omitted with little performance impact.

Glossary

Active circuit Any semiconductor component or product that uses transistors or FETs.

Admittance The reciprocal of impedance. It has a real part called a conductance and an imaginary part called a susceptance. The units of admittance are the siemens.

Admittance Smith chart The impedance Smith chart rotated 180°.

Complex notation The use of complex numbers to represent the 90° phase relationship between sine wave measures of current and voltage in inductors and capacitors. This notation extends to reflection and transmission coefficients. This notation is explained in Appendix A.

Conductance The reciprocal of resistance. The units are real siemens.

Dipole An antenna configuration. A full dipole is a length of the conductor fed from a center point. A half dipole is a conductor perpendicular to a conducting surface such as the earth.

Effective radiated power The power required to establish a field of uniform intensity in all directions.

Full dipole An antenna consisting of two in-line conductors driven by a balanced signal.

Half dipole An antenna consisting of a single conductor usually mounted perpendicular to a conducting surface.

Impedance The opposition to current flow in any network. It generally applies to sine waves. It is used as the ratio of electric to magnetic field intensities in space and along transmission lines. For a transmission line, it is usually a real number. In a network, it is usually a complex number. Impedance has a real component called resistance and a complex component called a reactance.

Impedance of free space 377 ohms. The ratio of the electric field intensity to the magnetic field intensity in space at a distance from the radiator (Section 5.1).

Interface distance The distance from the radiator where the wave impedance is approximately 377 ohms. This distance is $\lambda/2\pi$.

Near-field/far-field interface distance The distance from a radiator where the wave impedance E/H is equal to 377 ohms.

Radiation The release of electromagnetic energy that does not return to the generating source.

Reactance The opposition to current flow in capacitors and inductors. An inductor has a positive reactance and a capacitor a negative reactance. The unit of reactance is the ohm.

Siemens The unit of admittance. Admittance has a real part called conductance and an imaginary part called susceptance.

Smith chart A plot of the complex transmission coefficients τ for a terminated transmission line with a characteristic impedance of 1 ohm terminated in any impedance. The terminating impedances are found on orthogonal circles located inside a unit circle. The center of this circle is where there is full transmission and $\tau = 1$ (Section 5.4).

Solid angle Consider a circular cone with its tip at the center of a unit sphere. The surface area intersected by the cone is a measure of solid angle. The total surface area of a sphere is $4\pi r^2$. The full solid angle is 4π steradians.

Standing wave ratio The ratio of peak voltage to minimum voltage along a transmission line carrying a steady sine wave signal. Abbreviated SWR. This pattern is stationary along the transmission line. For an ideal transmission the SWR is unity (Section 5.2).

Stub A short section of transmission line. It can be shorted or open circuited at the far end.

Susceptance For sine waves. The reciprocal of reactance. The ratio of current to voltage for a capacitor or inductor when a voltage is applied. In circuit theory, the susceptance of a capacitor is a positive imaginary number of siemens. The susceptance of an inductor is a negative imaginary number of siemens.

SWR See standing wave ratio.

Wavelength The distance between positive peaks for a sine wave pattern.

Appendix A

Sine Waves in Circuits

A.1 Introduction

The material in this appendix is a fast review of linear circuit theory. This material is needed when applying the Smith chart to matching impedances. This theory describes the voltages and currents that result when sine waves at a fixed frequency are used to excite a network of linear components. Complex notation is used so that the phase relations in RLC networks can be analyzed as a function of frequency. The concepts of impedance, admittance, resistance, conductance, reactance, and susceptance are introduced. If the reader needs a slower pace, there are many texts available with problem sets. Working problems is an important part of learning this material and keeping it fresh.

A.2 Unit Circle and Sine Waves

On a simple grid of vertical and horizontal lines, label the center point as zero. This point is called the origin. Draw a circle with the center in mid page. This is called the unit circle. The horizontal line that goes through the origin is the x axis, the vertical line that goes through the origin is the y axis. The top of the circle is a y value of 1. The bottom is −1. The far left point has a y value of 0 and an x value of −1. Number the grid of lines so that every point on the circle can be given an x and y value from 0 to ±1.

Now draw a radial line from the origin to the circle at the point where $y = 0.5$. The radius is the hypotenuse of a triangle formed by the x and y distances. Note that any point on the circle must satisfy the relationship $x^2 + y^2 = 1$. This means that the x value here is 0.866 when the y value is 0.5. This point is often labeled $(x, y) = (0.5, 0.866)$ (see Figure A.1).

Label the angle between the radius and the x axis β. When the angle is increased to 45°, the value of x and y are both 0.707. When the radius is vertical

Fast Circuit Boards: Energy Management, First Edition. Ralph Morrison.

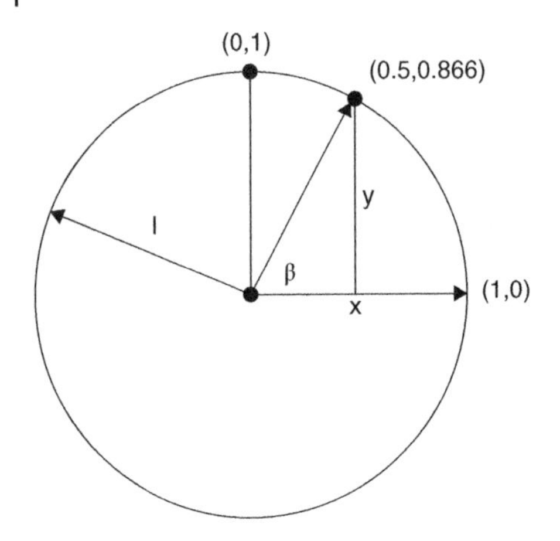

Figure A.1 The unit circle and the point (0.5, 0.866).

the value of x is 0 and the y value is 1.0. This point on the diagram is called (0, 1). The angle $\beta = 90°$.

Continue to increase the angle until the tip of the radius makes a full circle. In one revolution this angle is 360°. Two full cycles would result in an angle of 720°. When the angle is 150°, the y value returns to 0.5. When the angle is 270°, the y value is −1. This point is identified as (0, −1).

This vertical or y value in this diagram is called the sine of the angle. If the line is above the x axis the sine value is positive. If the line is below the x axis the sine value is negative. If the angle is labeled β, the equation that relates the y value to the angle β is

$$y = \sin\beta \tag{A1}$$

where sin is the abbreviation for sine. This equation is read "y equals sine of β." The value of y is always between +1 and −1. When β is 0° or 180°, the value of y is 0. This means that sin (180°) = 0. When $\beta = 90°$, sin (90°) = 1.

The sine function is a ratio of line lengths using the unit circle. It is one of six trigonometric functions. The x value in the unit circle example above is called the cosine function. It is abbreviated as cos β. In equation form we can write

$$x = \cos\beta. \tag{A2}$$

The value of the cosine function when $\beta = 0°$ is 1: when $\beta = 90°$ the value of cos β is 0. When $\beta = 180°$ the value of cos β is −1.

Now rotate the radial counterclockwise at a constant rate starting with the tip at the point (1, 0). On a second graph, plot the vertical position of the tip against time. The curve that results (shown in Figure A.2) is called a sine wave. If the radial starts at (0, 1) and rotates at a steady rate, a cosine wave is formed. Sine waves and cosine waves have the same shape but the peak values occur at a different time. The cosine wave is said to lead the sine wave in phase by 90°. By convention any wave with this shape is called a sine wave.

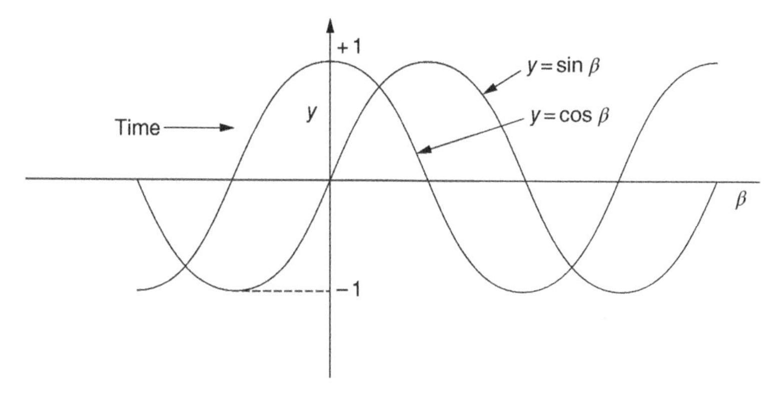

Figure A.2 A sine and cosine wave.

Problem Set A.2

1. Simplify $\sin 90° + \cos 180° - \sin 270°$.
2. A unit radial has an x value of 0.707. What is the y value?
3. At what two angles does $\sin \beta = 0.707$?
4. If $\sin \beta = 0.4$ what is $\cos \beta$?
5. What are the maximum and minimum values of the sine function?

A.3 Angles, Frequency, and rms

The measurement of angles dates well before man invented the calculus. Dividing the circle into 360 parts is a part of history. Some feel it would have made more sense to divide the circle into 100 parts rather than 360. In mathematics, the most useful way to measure angles involves the radius. Take a circle and wrap the radius along the circumference. The angle that is subtended by the radius is called a radian. Since 2π radians = 360°, we know that 1 radian is equal to 57.29°. This is not a nice even number. To get around the problem, electrical engineers measure angles in multiples of 2π. In other words, 2π radians equals 360°. An angle of $\pi/2$ radians is thus read as 90 electrical degrees. Most engineers refer to the angle as $\pi/2$ and omit the term radian. In the calculus the derivative of $\sin x$ is $\cos x$, but only if the angles are in radian units.

The sine of 90° is equal to 1. The convention that is in constant use in engineering is $\sin(90°) = 1$, which is written as $\sin(\pi/2) = 1$.

It is very important to realize that the sine function only applies to angles. In one cycle the angle changes 360° or 2π radians. To generate sine waves at a given frequency, say at 60 Hz which is really 60 cycles/s, the number of degrees that must pass in 1 second is 60 times 360° or 60 times 2π radians. By convention the angle in radians at a time t is simply

$$2\pi ft \tag{A3}$$

where f is the conventional frequency in hertz and t is time in seconds. Since the term $2\pi f$ occurs often in electronics it is, by convention, labeled the Greek letter ω. *The term* ω is called the radian frequency.

$$\omega = 2\pi f \tag{A4}$$

where f is frequency in hertz. The angle β in radians as a function of time is

$$\beta = \omega t. \tag{A5}$$

It is necessary to interpret all literature that describes sinusoidal responses as being in terms of radian frequency, not degrees frequency. A sine wave voltage would be described by

$$\mathrm{v} = \mathrm{V} \sin \omega t = \mathrm{V} \sin 2\pi ft \tag{A6}$$

where V is the peak voltage value, v is the value of voltage at time t, and ω is the radian frequency. Often, just the lower case letter v is used and the reader understands it is a sine wave. It is time consuming to write $\sin \omega t$ over and over. This understanding takes time to get used to. The sine waves in Figure A.2 are phase shifted by 90 electrical degrees. The cosine wave could have been written

$$\mathrm{v} = \mathrm{V} \sin(\omega t + \beta) \tag{A7}$$

where $\beta = \pi/2$.

The value V is called the peak value. The value v (in lower case) changes with time in a sinusoidal manner. This is also by convention. For current, the expression might be

$$i = I \sin \omega t. \tag{A8}$$

The voltage V or the current I are peak values. The value measured by a voltmeter is often the heating value. This is the peak sine value divided by the square root of 2 or 0.707 V. In most electronics the peak value is used, not the rms value. This is also by convention. In power engineering V would mean the rms or heating value, not the peak value. The utility power in a home is 120 V rms. This voltage has a peak value of 169.7 V and a peak-to-peak value of 339 V.

The current that flows when a voltage is applied to a resistor R is by Ohm's law V/R. The current can also be described by rotating a radial line. The rotating radius for the current points in the same direction as the voltage. The voltage and current are in phase. This means that when the voltage peaks the current also peaks. This rotating line is a vector as it has direction and intensity indicated by length.

Problem Set A.3

1. What is the value of $\sin(3\pi/2)$?
2. What is the value of $\cos(-\pi/2)$?
3. What is the radian frequency if the voltage vector rotates at twice per second?
4. At 100 Hz, how many degrees does the radial cover in 1 millisecond?
5. At 100 Hz, how many radians does the radial cover in 1 millisecond?
6. The voltage vector rotates once per microsecond. What is the frequency? What is the radian frequency?
7. A voltage has a peak value of 100 V. What is its rms value?
8. Write the equation for a sine wave at 1 kHz that leads another 1 kHz signal of equal amplitude by 30°.
9. A voltage $3\sin\omega t$ is added to a voltage $4\cos\omega t$. What is the sum?
10. A voltage $3\sin\omega t$ is added to a voltage $4\cos(\omega t + \pi/6)$. What is the sum?

A.4 The Reactance of an Inductor

In the circuit in Figure A.3, a sinusoidal voltage is connected to an inductor at the moment of maximum voltage. This timing avoids generating any transients. The current starts at zero and increases as long as the voltage is positive. When

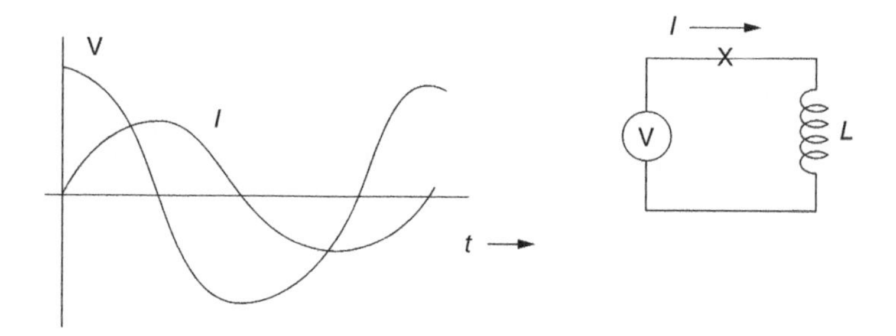

Figure A.3 Sine wave voltage and current for an inductor. The current lags the voltage by 90°.

the voltage is zero the current peaks. The current begins to decrease as the voltage goes negative. The current is zero when the voltage is at its negative peak. The current and voltage waveform are shown in Figure A.3.

The current lags the voltage by 90 electrical degrees. If $v = V \sin \omega t$ then the current I is equal to

$$I = -\frac{V \cos \omega t}{\omega L} \tag{A9}$$

where ω is the radian frequency, L is the inductance in henries, and V is the peak voltage.

This 90° relation between current and voltage occurs for every inductor handling sine wave signals. The ratio of peak voltage to peak current flow is called a reactance X_L measured in ohms. The peaks do not occur at the same time.

$$\frac{V}{I} = X_L = \omega L = 2\pi f L. \tag{A10}$$

If a steady current flows in an inductor that uses magnetic material, the material will saturate. This results in nonlinear behavior. Inductors that carry low-frequency currents should not use magnetic materials without an adequate air gap.

When inductors are paralleled the total current that flows is given by

$$I = \frac{V}{L_T} = \frac{V}{L_1} + \frac{V}{L_2}. \tag{A11}$$

Eliminating V, it is easy to see that parallel inductors add using reciprocals.

$$\frac{1}{L_T} = \frac{1}{L_1} + \frac{1}{L_2}. \tag{A12}$$

Problem Set A.4

1 What is the reactance of a 1 mH inductor at 10 kHz?
2 What is the current flow in a 1 μH inductor at 2 MHz if the voltage is 1 V peak?
3 What is the inductance of a 10 μH inductor in parallel with an 8 μH inductor?

A.5 The Reactance of a Capacitor

In the circuit of Figure A.4, a sinusoidal voltage is applied to a capacitor at the moment the voltage starts going positive. This timing avoids generating any transients.

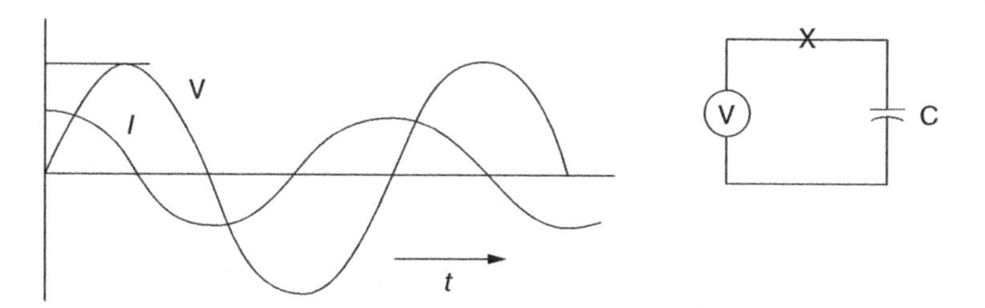

Figure A.4 Sine wave voltage and current for a capacitor. The voltage leads the current by 90°.

The current that flows depends on the rate of change of voltage which is maximum at the moment of contact. The current leads the voltage by 90 electrical degrees and is equal to

$$i = +\omega C V \cos \omega t \tag{A13}$$

where ω is the radian frequency, C is the capacitance in farads, and V is the peak voltage. This 90° relationship between current and voltage occurs for every capacitor when there are sine waves. The ratio of peak voltage to peak current is called a reactance X_C measured in ohms. The reactance is

$$X_C = \frac{V}{I} = \frac{1}{\omega C} = \frac{1}{2\pi f C}. \tag{A14}$$

When capacitors are placed in parallel their capacitances add directly.

When capacitors are put in series their reactances add directly. Using Equation A12 the reactance of series capacitors is

$$X = \frac{1}{\omega C_1} + \frac{1}{\omega C_2} = \frac{1}{\omega C_T}. \tag{A15}$$

Eliminating ω it is easy to see that series capacitances add by using reciprocals.

$$\frac{1}{C_T} = \frac{1}{C_1} + \frac{1}{C_2}. \tag{A16}$$

Problem Set A.5

1 What is the reactance of a 0.001 μF capacitor at 100 MHz?
2 What is the parallel capacitance of 1 nF and 200 pF?
3 What is the capacitance of 0.001 μF and 0.002 μF in series?
4 What capacitance has a reactance of 50 ohms at 400 MHz?

A.6 An Inductor and a Resistor in Series

Assume a current is flowing in a series resistor and inductor. The voltage across the inductor peaks 90 electrical degrees before the voltage across the resistor. If the voltage across the resistor is $IR\sin\omega t$, then the voltage across the inductor is $I\omega L\cos\omega t$. The sum of these voltages is

$$IR\sin\omega t + I\omega L\cos\omega t = \mathrm{V}\sin(\omega t + \phi). \tag{A17}$$

Expand the right term, then equate the coefficients of $\sin\omega t$ and $\cos\omega t$. It is not difficult to show that

$$\mathrm{V} = I\left(R^2 + \omega^2 L^2\right)^{½} \tag{A18}$$

and

$$\phi = \tan^{-1}\frac{\omega L}{R}. \tag{A19}$$

The ratio of V/I is called an impedance and is usually assigned the letter Z. In this case, the magnitude of the impedance in ohms of a resistor in series with an inductor is

$$|Z| = \left(R^2 + \omega^2 L^2\right)^{½}. \tag{A20}$$

Equations A19 and A20 are exactly the same as assigning a real value to a horizontal vector representing the voltage across the resistor and assigning an imaginary value to a vertical vector representing the voltage across the inductor. Using complex notation, the series RL circuit has an opposition to current flow equal to $Z = R + j\omega L$. The factor j is equal to the square root of -1. These vectors are shown in Figure A.5.

The impedance Z in ohms using complex notations is

$$Z = R + j\omega L. \tag{A21}$$

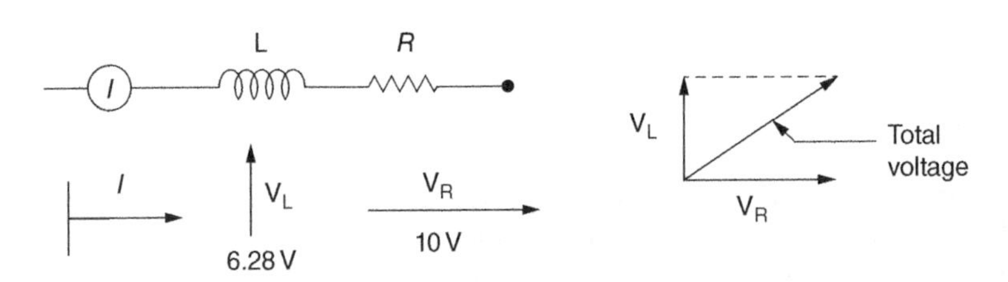

Figure A.5 The vectors representing the voltages in a series *RL* circuit.

The reactance X_L in ohms of the inductor is

$$X_L = j\omega L. \tag{A22}$$

A 10-ohm resistor in series with a 1-H inductor at 1 Hz has an impedance of (10 + 6.28*j*) ohms. This means that if a sine wave current of 1 A were to flow in this impedance, the voltage across the resistor would be a 10-V sine wave and the voltage across the inductor would be a 6.28-V cosine wave. The total voltage is $(10^2 + 6.28^2)^{½}$ V = 11.8 V. This is a sine wave voltage leading the voltage across the resistor by 32.3°.

Problem Set A.6

1 An inductor of 10 μH is in series with 50 ohms. What is the total impedance at 20 MHz?
2 An inductance has a reactance of 40 ohms at 50 MHz. What is the inductance?
3 An inductor of 3 μH is in series with 100 ohms. What is the phase angle between the current and voltage at 10 MHz?
4 At what frequency will an inductor of 20 nH have a reactance of 100 ohms?

A.7 A Capacitor and a Resistor in Series

When a resistor and a capacitor are placed in series, a current generates a voltage across the capacitor that peaks 90 electrical degrees after the voltage peaks across the resistor. If the voltage across the resistor is $IR \sin \omega t$, then the voltage across the capacitor is $-I\omega L \cos \omega t$. The sum of these voltages is

$$\mathrm{V} \sin(\omega t + \phi) = IR \sin \omega t - \frac{I}{\omega C} \cos \omega t. \tag{A23}$$

Expand the left term and equate coefficients of $\sin \omega t$ and $\cos \omega t$. The voltage V is

$$\mathrm{V} = I\left(R^2 + \frac{1}{\omega^2 C^2}\right)^{1/2} = IZ \tag{A24}$$

and

$$\phi = \tan^{-1} \omega CR. \tag{A25}$$

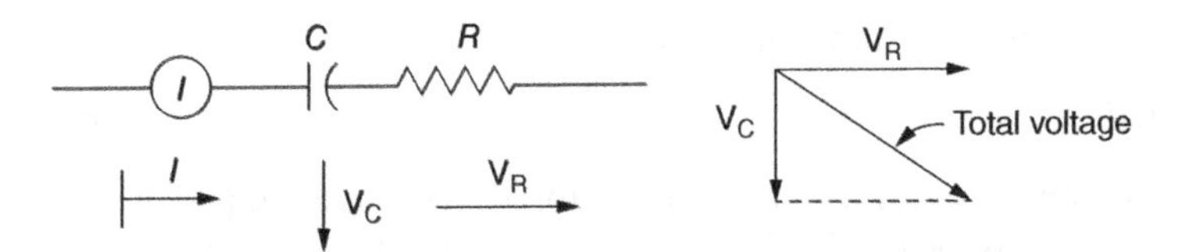

Figure A.6 The vectors representing the voltages in a series *RC* circuit.

In complex notation, the opposition to current flow for a resistor and capacitor in series is an impedance Z where

$$Z = R - \frac{j}{\omega C}. \tag{A26}$$

The magnitude of the impedance in ohms is

$$|Z| = \left(R^2 + X_C^2\right)^{1/2} \tag{A27}$$

and the reactance of the capacitor is equal to

$$X_C = \frac{-j}{\omega C}. \tag{A28}$$

These vectors are shown in Figure A.6.

A 3-ohm resistor in series with a ½ farad capacitor at 2 Hz has an impedance of 3–0.159j ohms. What this means is that if the voltage across the resistor is a 3-V sine wave, the voltage across the capacitor is a delayed cosine wave of 0.159 V.

Problem Set A.7

1 At what frequency will a capacitor of 100 pF and a series resistor of 100 ohms have an impedance of 200 ohms?
2 In problem 1, at what frequency will the reactance double?
3 The series impedance in problem 1 is 300 ohms. What is the phase angle in degrees between the voltage and the current flow?

A.8 The Arithmetic of Complex Numbers

It is very convenient to use complex numbers to represent voltages, currents, impedances, and admittances. In a series circuit the voltages across components do not add directly as the peaks of voltages do not occur at the same time. In a parallel circuit the peaks of current do not occur at the same time. Complex numbers are the simplest way to solve these circuit problems. Here are the rules for addition, multiplication, and division of complex numbers.

Addition. Add the real and imaginary terms separately.

$$(2+3j)+(3-4j)=(-1+7j)$$

Multiplication. Use standard rules for multiplication where $j^2=-1$, $j^3=-j$ and $j^4=1$.

$$(2+3j)(3-4j)=(18-j)$$

Division requires the use of conjugates. The conjugate of $(2+3j)$ is $(2-3j)$.

To divide by a complex number, multiply the numerator and denominator by the conjugate of the denominator. Here are two examples:

$$\frac{1}{2+3j}=\frac{2-3j}{(2-3j)(2+3j)}=\frac{2-3j}{13}=2/13-3j/13$$

$$\frac{(1+2j)}{(2+3j)}=\frac{(1+2j)(2-3j)}{(2+3j)(2-3j)}=\frac{8-j}{13}=8/13-j/13$$

A.9 Resistance, Conductance, Susceptance, Reactance, Admittance, and Impedance

Resistor inductors and capacitors have the ability to oppose current flow. The unit that is used for resistance is the ohm abbreviated as Ω. The opposition to current flow by capacitors and inductors is called reactance and uses the symbol X. An impedance Z is a combination of resistance and reactance. These same components have the ability to accept current flow. The terms that are used are conductivity for a resistor and susceptance for a capacitor or inductor. The unit is the siemen, abbreviated as S.[1] A combination of a conductance and a suseptance is called an admittance and uses the letter Y. In each case the measure of current acceptance is the reciprocal of current opposition. For example, a resistor of 10Ω has a conductance of 0.1 siemens (S). It is the same resistor with two different measures. A capacitor has a reactance of $-2j$ Ω and a susceptance of $0.5j$S. It is the same capacitor. An inductor has a reactance of $5j$ Ω and a susceptance of $-0.2j$S. It is the same inductor. Any impedance can be described as a susceptance and visa versa. The ratio of

1 The siemen (S) has replaced the mho as the unit of admittance. Ernst Werner Siemens was a German born inventer.

voltage to current is fixed at any one frequency for a given network. It takes practice to get used to the language that is used.

When resistors are paralleled their conductances add. A 5-Ω resistor in parallel with a 2-Ω resistor is the sum of two conductances namely $1/5 + 1/2 = 7/10$ S. To convert this conductance back to a resistance take the reciprocal which is $10/7\ \Omega$.

When inductors are paralleled the same principle is used. Each inductor is described as a susceptance and these susceptances are added together. The result can be converted to a reactance by taking the reciprocal. To parallel an inductor of $2j\,\Omega$ with an inductor of $4j\,\Omega$ their susceptances are $-0.5j$ and $-0.25j$ S. Their sum is $-0.75j$ S. The reciprocal is a reactance of $+1.33j\,\Omega$.

When capacitors are paralleled each capacitor is described as a susceptance and the susceptances are added together. The result can be converted to a reactance by taking the reciprocal. To parallel a capacitor of $-4j$ ohms with a capacitor of $-5j$ ohms, their susceptances are $+0.25j$ and $+0.2j$ S. Their sum is $+0.45j$ S. To convert to a reactance the reciprocal is $-2.22j$ ohms.

The admittance Y of a resistor of $5\,\Omega$ in parallel with an inductor of $4j\,\Omega$ is found by converting the resistance to a conductance of 0.20 S and the inductor to a susceptance of $-0.25j$ S. The admittance is the sum $Y = (0.20 - 0.25j)$ S. To covert this to an impedance, take the reciprocal which is $Z = (1.957 + 2.439j)\,\Omega$.

The admittance Y of a resistor of $5\,\Omega$ in parallel with a capacitor of $-2j\,\Omega$ is found by converting the resistance to a conductance of 0.2 S and the capacitor to a susceptance of $+0.5j$ S. The admittance is the sum $Y = (0.2 + 0.5j)$ S. The impedance of this parallel circuit is the reciprocal of Y which is $Z = (0.69 - 1.76j)\,\Omega$.

Problem Set A.9

In these problems use complex notation.

1 Two capacitors have a reactance of $-2j$ and $-3j$ ohms. What is the series reactance?
2 Two inductors have a reactance $0.5j$ and $0.6j$ ohms. What is the parallel reactance?
3 A resistor of 2 ohms is in series with an inductance of $3j$-ohm reactance. What is the admittance in siemens?
4 A resistor of 3 ohms is in parallel with a capacitor with a reactance of $-4j$ ohms. What is the impedance in ohms?
5 In problem 4, what is the admittance in siemens?
6 A capacitor with a reactance of $2j$ is in parallel with an inductance having a susceptance of $-2j$. What is the impedance?

7 An impedance of $2+3j$ is in series with an admission of $1-2j$. What is the impedance?

8 A susceptance of $3+2j\text{S}$ is in series with an impedance of $2-3j$ ohms. What is the total impedance?

N.B.

Every network of RLC components can be represented as either an impedance or an admittance. Impedances can only be added or subtracted from impedances. Admittances can only be added or subtracted from admittances.

Admittances are reciprocals of impedances.

Susceptances are reciprocals of reactances.

Conductances are reciprocals of resistances.

In an impedance, the real part is a resistance and the imaginary part is a reactance. A positive reactance is an inductor and a negative reactance is a capacitor.

In an admittance the real part is a conductance and the imaginary part is a susceptance. A positive susceptance is a capacitor and a negative susceptance is an inductor.

A.10 Resonance

The reactance of an inductor is given in Equation A18. The reactance of a capacitor is give in Equation A22. The impedance of the two elements in series is

$$Z = j\omega L - \frac{j}{\omega C}. \tag{A29}$$

The frequency ω where $Z=0$ is called the series resonant frequency. The frequency is

$$\omega = \frac{1}{(LC)^{1/2}}. \tag{A30}$$

If there is a series resistance the impedance at the resonant frequency is this resistance.

The susceptance of an inductor is $-j/\omega L$. The susceptance of a capacitor is $j\omega C$.

The frequency where the susceptances adds to zero is given by Equation A26. This frequency is called the parallel resonant frequency. This circuit is sometimes called a tank circuit. Ideally the impedance of a parallel resonant circuit is infinite at the resonant frequency.

A.11 Answers to Problems

Problem Set A.2

1. Sin 90° = 1, cos 180° = −1, sin 270° − 1. The sum is 1.
2. 0.707.
3. 45° and 135°.
4. Cos $\beta = 0.917$.
5. The limits are +1 and −1.

Problem Set A.3

1. Sin $3\pi/2 = -1$
2. Cos $-\pi/2 = 0$.
3. 4π.
4. 36°.
5. $\pi/5$.
6. $f = 1$ MHz, $2\pi \times 10^6$ radians per second.
7. 70.7 V
8. Sin$(2000\pi t + \pi/6)$.
9. $5 \sin(\omega t + 53°)$.
10. Sin $\omega t + 3.46 \cos \omega t$.

Problem Set A.4

1. 20π.
2. $1/4\pi$ peak.
3. 4.44 μH

Problem Set A.5

1. 1.59 ohms.
2. 1.2 nF.
3. 0.666 nF.
4. 398 pF.

Problem Set A.6

1. 1.25 kΩ.
2. 0.127 μH.
3. 62°.
4. 796 MHz.

Problem Set A.7

1 9.19 MHz.
2 18.38 MHz.
3 20.7°.

Problem Set A.8

1 $-5j$.
2 $1.1j$.
3 $(2-3j)/13$.
4 $12(4-3j)/25$.
5 $1/3+j/4$.
6 $12(4-3j)/25$.
7 $2.2+3.4j$.
8 $29/13-41j/13$.

Appendix B

Square-Wave Frequency Spectrum

B.1 Introduction

One of the most powerful tools in circuit analysis is the Fourier analysis. Signals that enter a circuit are often repetitive but not sinusoidal. Typical repetitive signals are square waves, triangle waves, trapezoids, or pulses. Square waves can have different leading and falling edges. A Fourier analysis provides an infinite series of sine waves with different amplitudes that when added together create the original waveform. The response to a repetitive waveform is found by summing the responses to all the individual sine waves (harmonics). In most practical applications, the response to the first three harmonics is sufficient to characterize the amplitude of the waveform. The assumption that must be made is that the circuit being analyzed is linear. This means that the system does not generate new harmonics.

B.2 Ideal Square Waves

Figure B.1 shows how an ideal square wave voltage is built up from sinusoids. If the square wave were shifted in time by 90 electrical degrees, the harmonics would all be cosines.

The graphs that follow show the nature of the spectrum for square waves. Practical logic signals are not fully repetitive, so the harmonic content shown in the figures is approximate. Note that the amplitude A is one-half the peak-to-peak value. For 5-V logic, A is 2.5 V. The envelope of amplitudes is presented only as a guide. There is no energy content except at the harmonic frequencies.

The amplitudes of the harmonics of an ideal square wave on both a linear and a logarithmic scale are shown in Figure B.2. The harmonics that make up an ideal square wave have amplitudes that are inversely proportional to

Fast Circuit Boards: Energy Management, First Edition. Ralph Morrison.

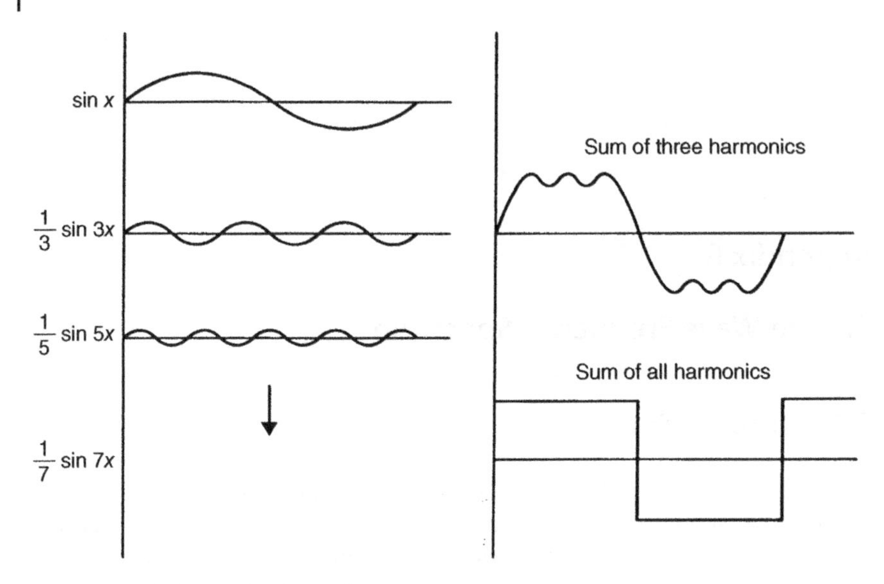

Figure B.1 The harmonics that make up a square wave.

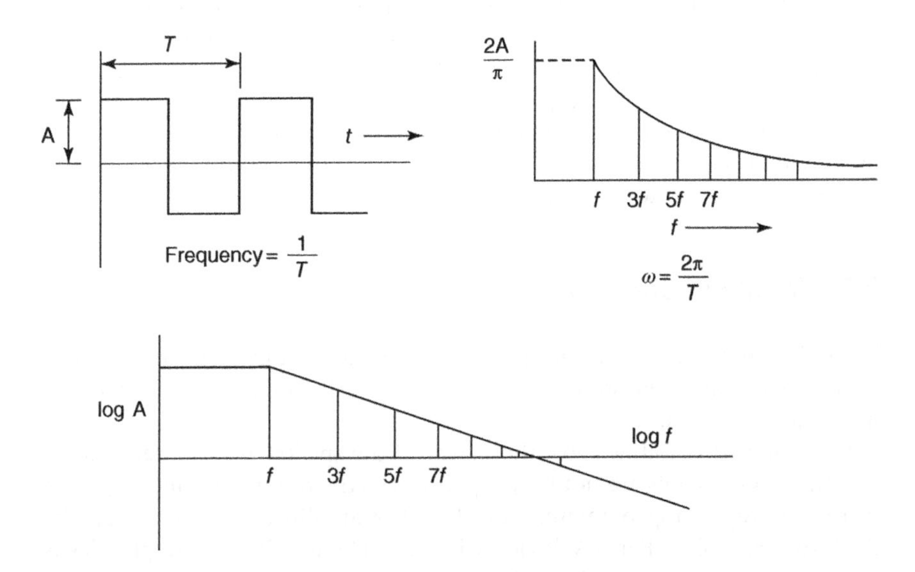

Figure B.2 The harmonics that make up a square wave plotted linearly and on a logarithmic scale.

frequency. On a logarithmic scale the amplitudes lie along a straight line with a negative slope of one. The fundamental frequency is simply $1/t$, where t is the time of one full cycle. The amplitude of the fundamental or first harmonic sine wave voltage is

$$V = \frac{2A}{\pi} \sin \omega t \tag{B1}$$

where

$$\omega = \frac{2\pi}{t}. \tag{a2}$$

The amplitude of the nth harmonic is

$$V_n = \frac{2A}{n\pi} \sin \omega_n t. \tag{B3}$$

B.3 Square Waves with a Rise Time

A repetitive square wave with a symmetric rise and fall time can be represented by a series of sine waves. The amplitudes of the harmonics determined in a Fourier analysis vary in a complex way. Figure B.3 shows that the harmonics have amplitudes that are contained under a logarithmic envelope with two slopes. These slopes intersect at frequencies that are called corner frequencies. The first or lower corner frequency is at $f = 1/\pi\tau$ where τ is the time between zero crossings. The amplitude of the harmonics above this frequency falls off

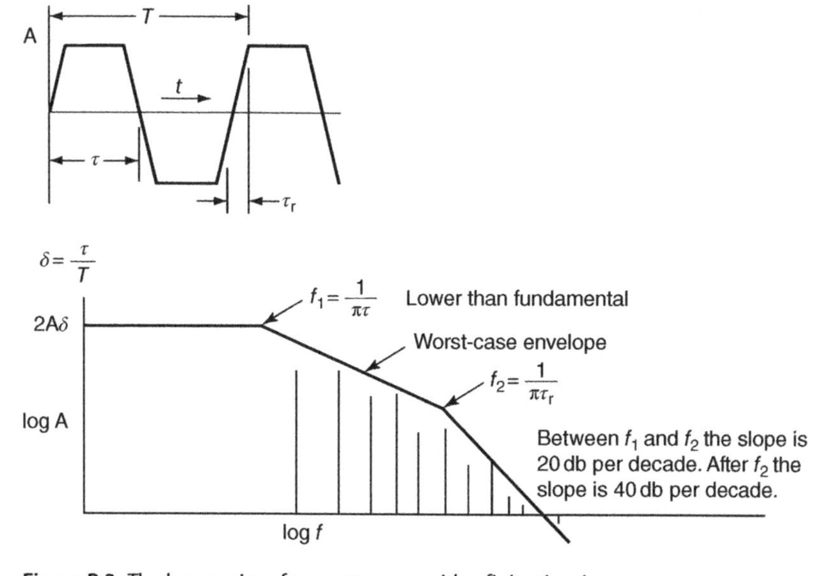

Figure B.3 The harmonics of a square wave with a finite rise time.

proportional to frequency or 20 dB per decade. The second or upper corner frequency is at $f = 1/\pi\tau_R$ where τ_R is the rise time. Above this upper corner frequency the harmonic amplitudes fall off at the square of frequency or 40 dB per decade.

The two corner frequencies have significance when considered in terms of interference coupling. In general, coupling processes increase proportional to frequency. This would cancel the attenuation of harmonics in the frequency range $1/\pi\tau$ to $1/\pi\tau_R$. This would imply that the interfering signal is effective out to the second corner frequency. Above $1/\pi\tau_R$, coupling falls off proportional to frequency. The impact of interference is thus related to its upper corner frequency.

Appendix C

The Decibel

There are many parameters in engineering that extend in value over many decades. As an example, useful voltages extend from well below a microvolt to megavolts. That is over 12 orders of magnitude. Often it is convenient to consider a logarithmic scale rather than a linear scale to discuss parameters. In the early days of telephony, engineers needed a logarithmic scale for discussing sound levels. Noise levels, for example, were millivolts while signal levels were volts. The power ratio here is a million to one. It was convenient to use a logarithmic scale, where a just discernable change in loudness was 1 unit.[1] This unit had to work for noise as well as voice signals. It turned out that the logarithm of the ratio of power from two signals when multiplied by 10 was just such a scale. It worked at all signal levels. The unit was called the decibel to honor Mr. Bell, the telephone inventor. The definition of the decibel is

$$\mathrm{dB} = 10\log_{10}\left(\frac{P_1}{P_2}\right) \tag{C1}$$

where P_1 and P_2 are power levels. If the signal levels are measured as voltage on one resistor, then the ratio can be written as

$$\mathrm{dB} = 10\log_{10}\frac{(V_1)^2 R}{(V_2)^2 R} = 20\log_{10}\frac{(V_1)}{(V_2)}. \tag{C2}$$

The decibel is abbreviated as dB and in conversation it is pronounced "deebee." For those that use this measure on a regular basis, they know that 6 dB is a factor of 2, 20 dB is a factor of 10, and that negative decibel figures imply division. For example, −6 dB means a factor of 0.5 or divide by 2, and −20 dB means a factor of 0.1 or divide by 10. The factor 50 is 100/2 and in decibel language this is 40 dB – 6 dB or 34 dB. It takes usage to recognize this language.

1 The logarithm of a number to the base 10 is the exponent of 10 that equals that number. For example, the logarithm of 100 is 2 because $10^2 = 100$. The logarithm of 2 is 0.30103 because $10^{0.30103} = 2$. Note that $20\log_{10} 2$ is usually rounded off to 6 dB.

Fast Circuit Boards: Energy Management, First Edition. Ralph Morrison.

In field measurement, the units can be volts, current, the H or E field, and watts. In fact, the decibel scale can be applied to nonelectrical parameters such as meters. This means that units must usually be associated with a decibel statement. It is a standard practice to refer to 20 dBV, which means 10 V. A total of 6 dBV means 2 V. It is important to know that 0 dB means a factor of 1. There is no decibel representation of 0 V as the logarithm of 0 is $-\infty$.

The parameters P_2 or V_2 are called reference parameters. They might be units of watts or volts, milliwatts, or millivolts as well as megawatts or megavolts. If the reference parameter is a millivolt, then 20 dBmV means 10 mV. The reference parameter must be stated or the decibel statement has no meaning. There are a group of abbreviations that have become standard. Again it takes usage to become familiar with this language. Here are a few standard abbreviations.

dBV	dB volts	The reference is 1 V
dBmV	dB millivolts	The reference is 1 mV
dBmW	dB milliwatts	The reference is 1 mW
dBμV	dB microvolts	The reference is 1 μV
dBV/MHz	dB volts per megahertz	The reference is 1 V/MHz
dBrn	dB reference noise	Used in telephony. The noise reference level is −90 dBW

The term decibel usually implies a power ratio. When the units are volts or amperes, Equation C2 is used. When the units are power, Equation C1 is used. When the units are a parameter such as ohms or meters, it is obvious that power does not apply. In these cases, the term decibel represents a logarithmic scale of 20 log 10 A/B. Many other disciplines that need logarithmic scales have adapted the decibel for their use. An example is a specification for street paving. Roughness is described in terms of decibel inches. The reference parameter is 1 inch and this has little to do with power. Equation C2 is used.

Appendix D

Abbreviations and Acronyms

New acronyms appear in the literature each year. As time goes by, many of these acronyms fall into disuse. In a few cases, meanings may differ depending on the industry where they originate. It is possible for one abbreviation to have several meanings. When this happens, the reader must rely on the intent of the author. As an example, pH can mean hydrogen ion concentration or picohenry. The correct meaning should be obvious in context.

Authors often coin an acronym in an article when a word group is repeated many times. Sometimes, an acronym is used in one company, but it is not used by the industry. These acronyms will not appear in this list.

The following list includes the common abbreviations used in circuit board design, mathematics, electrical engineering, and physics. Fortunately, acronyms that survive rarely conflict with scientific or engineering abbreviations.

Abbreviations that do not conform to present day usage or good practice are not listed. Here are a few of the problem areas. The first letter in many engineering measurement terms is an "m." In an abbreviation, "m" could stand for milli, micro, or mega. It is good practice to use capital M for mega, small m for milli, and the Greek μ for micro. The letter m can also stand for meter, milliwatt, and mile. In the abbreviation dBm, the m means milliwatt. If m means meter, it can usually be inferred by context.

In an old copy of *The Radio Engineers Handbook* by Terman, the term mc is used as the abbreviation for megacycle. Today, the accepted abbreviation is MHz and it is read megahertz. We have come a long way in standardizing abbreviations.

Proper names used in engineering abbreviations require a capital letter. As an example, the A in mA stands for Andre Marie Ampere, the French physicist. When the word ampere is used in a sentence, it is not capitalized. "ma" is not found in this list.

This list contains most of the common abbreviations used in this book and in circuit board design. It is not a complete list. It shows how extensively we use abbreviations and acronyms. It is hard to remember them all. If you do not find an abbreviation, the internet will usually provide some help.

Fast Circuit Boards: Energy Management, First Edition. Ralph Morrison.

A	Ampere, the unit of current
ac	Alternating current. Usually a sine wave of voltage or current
A/D	Analog-to-digital
am	Amplitude modulation
amp	Ampere (not recommended)
ANSI	American National Standards Institute
ASCII	American standard code for information interchange
ASIC	Applications specific integrated circuit
ASSP	Application-specific standard product
ASTTL	Advanced Schottky transistor—transistor logic
ATE	Automatic test equipment
B	Magnetic induction field, measured in teslas
BGA	Ball grid array
BH	*B* and *H* fields in a magnetic material. Hysteresis curves
BOM	Bill of materials
b/s	Bits per second
BST	Barium–strontium–titanate (ceramic)
BTL	Bipolar transistor logic
BTU	British thermal unit
BW	Bandwidth
c	Velocity of light
C	Capacitor, capacitance
C	Coulomb. The unit of charge
C_{12}	A mutual capacitance
CAD	Computer-aided design
CAM	Computer-aided manufacturing
cc	Cubic centimeter
CFFT	Complex fast Fourier transform
CISPER A and B	EMI standards that are set by the EU
cm	centimeter
CMOS	Complimentary metal oxide semiconductor
CMR	Common-mode rejection
CMRR	Common-mode rejection ratio (analog)
cos	Cosine
cot	Cotangent
CPC	Computer power center
cps	Cycles per second. Not in common use
CRT	Cathode ray tube
CSA	Canadian Standards Association
csc	Cosecant
Cu	Copper
D	Displacement field
D/A	Digital to analog

dB	Decibel. 10 log P_1/P_2 or 20 log V_1/V_2. P_2 or V_2 are reference levels
dBA	Decibel-amperes
dBm	Decibel-milliwatts
dBV	Decibel-volts
dBμV	Decibel-microvolts
dc	Direct current. A steady value describing volts, current, or field strength
D_f	Dissipation factor
DFF	D flip flop
DFM	Design for manufacturability
DFT	1. Design for test
	2. Discrete Fourier transform
DIP	Dual inline package
D_k	Dielectric constant
DRAM	Dynamic random access memory
DTL	Decoupling transmission line
e	2.71828. Base of natural logarithms
E	Electric field strength. Basic unit is volts per meter
E	Energy in joules
ECAD	Electronic computer-aided design
ECL	Emitter-coupled logic
EDA	Electronic design automation
EE	Electrical engineer
EIA	Electronics Industry Association
EM	Electromagnetic
EMC	Electromagnetic compatibility
emf	Electromotive force, voltage
emi	Electromagnetic interference
ENIG	Electroless nickel/immersion gold
EPROM	Electrically programmable read only memory
ESD	Electrostatic discharge
ESL	Equivalent series inductance
ESR	Equivalent series resistance
EU	European Union
f	Frequency
F	Farad, the unit of capacitance
FB	Feedback
FCC	Federal Communications Commission
FET	Field effect transistor
FFT	Fast Fourier transforms
FIFO	First in first out
FILO	First in last out
fm	Frequency modulation
FP	Field programmable

FPGA	Field programmable gate array
FR-4	Flame resistant class 4 circuit board laminate. Glass epoxy
FTTL	Fast TTL
g	Gram
G	Gauss, 10^{-4} teslas
GaAs	Gallium arsenide
Gb/s	Gigabits per second
GHz	Gigahertz, 10^9 Hz
GND	Ground
GTL	Gunning transistor logic
H	Magnetic field strength, basic unit is amperes per meter
H	Henry, unit of inductance
HASL	Hot air solder leveling
HDI	High density interconnect
HF	High frequency
HSTL	High speed transceiver logic
Hz	Hertz (frequency). 1 Hz = 1 cycle per second
i	1. Varying current
	2. Square root of −1, used in most mathematics texts
IBIS	I/O buffer information specification
IC	Integrated circuit
id	Inside diameter
IEEE	Institute of Electrical and Electronic Engineers
I/O	Input/output
IP	Intellectual property
IPC	Institute of Printed Circuits
IR	Infrared
ISDN	Integrated Services Digital Network
J	Joule
j	Square root of −1
JEDEC	Joint Electronics Device Engineering Council
JFET	Junction FET
k	Kilo, 10^3
KB	Keyboard
kb/s	Kilobytes per second
kg	Kilogram
kHz	Kilohertz, 10^3 Hz
km	Kilometer, 10^3 m
kV	Kilovolt, 10^3 V
kW	Kilowatt, 10^3 W
kΩ	Kiloohms, 10^3 ohm
L	Inductor, inductance
L_{12}	Mutual inductance

LAN	Local area network
LCC	Leaded chip carrier
LED	Light emitting diode
LICA	Low-inductance capacitor array
ln	Natural logarithm (base e)
log	Logarithm (base 10)
LRC	Inductor/resistor/capacitor
LVCMOS	Low-voltage CMOS
LVDS	Low-voltage differential signaling
m	1. Meter 2. Milli
M	Mega, 10^6 (M mean 1000 in Roman numerals but not in engineering)
mA	Milliampere, 10^{-3} A
Mb/s	Mega bits per second
MCAD	Mechanical computer-aided design
MCM	Multichip module
mH	Millihenry
MHz	Megahertz, 10^6 Hz
mil	0.001 inches
Mil	Military
Mil Std	Military standard
mJ	Millijoule
mm	Millimeter
mmf	Magneto motive force
MOS	Metal oxide semiconductor
MOSFET	Metal oxide semiconductor field effect transistor
ms or msec	Millisecond, 10^{-3} s
mV	Millivolt, 10^{-3} V
mW	Milliwatt, 10^{-3} W
mΩ	Milliohm, 10^{-3} ohm
MΩ	Megaohm, 10^6 ohm
N	Newton
n	Nano, 10^{-9}
nA	Nanoampere, 10^{-9} A
NEC	National Electrical Code
NEMA	National Electrical Manufacturing Association
nF	Nanofarad, 1000 pF, 0.001 μF, 10^{-9} F
nH	Nanohenry
nm	Nanometer
npn	Transistor made from a p-doped semiconductor layer between two *n* layers
ns	Nanosecond, 10^{-9} s

OC Optical carrier
od Outside diameter
OEM Original equipment manufacturer
p Pico, 10^{-12}
PC, pc 1. Printed circuit
2. Personal computer
PCA Printed circuit assembly
PCB Printed circuit board
PCI Personal computer interface
PE Professional engineer
PECL Positive ECL
p_f Power factor
pF Picofarad, 10^{-12} F
pH Picohenry, 10^{-12} H
PLL Phase-locked loop
pn Junction of positive and negative doped semiconductor, diode
pnp Transistor, n-doped semiconductor between two p layers
ps Picosecond, 10^{-12} s
PTFE Polytetrafluoroethylene (dielectric)
PWB Printed wiring board
PWR Power
Q charge in coulombs
QA Quality assurance
QFP Quad flat pack
R Resistor, resistance
RAM Random access memory
RC Resistor/capacitor
rf Radio frequency
rfi Radio frequency interference, general interference
rms Root mean square
RoHM Reduction of hazardous material (lead free)
ROM Read only memory
rpm Revolutions per minute
rps Revolutions per second
RTI Referred to input
RTO Referred to output
s Second
S Siemens. The unit of admittance. The real part is conductance, the imaginary part is susceptance
SCR Silicon controlled rectifier
sec 1. Second (not recommended)
2. Secant
Si Silicon
SI Signal integrity

SIP Single inline package
SMD Surface mounted device
SMT Surface mount transistor
SPICE Special program for integrated circuit emulation
SRAM Static random access memory
SWR Standing wave ratio
t Time
T 1. tesla, the unit of magnetic induction
2. Temperature
tan Tangent
TBD To be determined
TC Temperature coefficient
TDR Time domain reflectometer
TE Transverse electric (wave)
TEL 1. Transitional electrical length. The distance a wave travels in a rise time
2. Telephone
TELCO Telephone company
T_g Resin transition temperature in laminates
TIA Telecommunications Industry Association
TL Transmission line
TM Transverse magnetic (wave)
TQFP Thin quad flat pack
TTL Transistor– transistor logic
TV Television
UHF Ultrahigh frequency
UL Underwriters Laboratories
USB Universal serial bus
UTP Unshielded twisted pair
v Varying voltage
V Volt
V Volume usually in meters cubed
VA Volt amperes
VAR Volt amperes reactive
Vcc Positive voltage on a circuit board
VCR Voltage controlled rectifier
Vdd Positive voltage on a circuit board
VME Type of bus and hardware protocol
VOH_{max} Highest loaded voltage for a logic 1 of a driver
VOH_{min} Lowest loaded voltage for a logic 1 of a driver
VOL_{max} Highest loaded voltage for a logic 0 of a driver
Vss The ground or most negative voltage on a circuit board
VSW Voltage standing wave
W Watt

W	Work
wan	Wide area network
X	Reactance in ohms
Y	Admittance in siemens. The real part of admittance is conductance; the imaginary part is susceptance. Plus imaginary susceptance is associated with shunt capacitance. Negative imaginary susceptance is associated with shunt inductance
Z	Impedance in ohms. The real part of impedance is resistance; the imaginary part is reactance. Plus imaginary reactance is associated with series inductance. Negative imaginary reactance is associated with series capacitance.
β	Current gain for a transistor
ε_R	Relative dielectric constant, permittivity
θ	Angle, phase, or phase angle
λ	Wavelength
μ_R	Relative permeability
μ_0	Permeability of free space
μ	Micro, 10^{-6}
μA	Microampere, 10^{-6} A
μF	Microfarad, 10^{-6} F
μH	Microhenry
μm	Micrometer, 10^{-6} m
μs	Microsecond, 10^{-6} s
μV	Microvolt, 10^{-6} V
π	3.14159
ρ	1. Resistivity 2. Reflection coefficient
σ	Conductivity
τ	Transmission coefficient
τ_r	Rise or fall time
φ	Angle, phase, or phase angle
ω	Radian frequency, $2\pi f$
Ω	ohm

Index

Fast Circuit Boards: Energy Management, First Edition. Ralph Morrison.

Printed and bound by CPI Group (UK) Ltd, Croydon, CR0 4YY
16/04/2025
14658595-0001